Photography

Foundations for Art & Design

2

Photography

Foundations for Art & Design

A guide to creative photography

Mark Galer

Focal Press
An imprint of Butterworth-Heinemann Ltd
Linacre House, Jordan Hill, Oxford OX2 8DP

ℛ A member of the Reed Elsevier plc group

OXFORD   LONDON   BOSTON
MUNICH   NEW DELHI   SINGAPORE   SYDNEY
TOKYO   TORONTO   WELLINGTON

First published 1995

British Library Cataloguing in Publication Data
A catalogue record for this book is available from the British Library

ISBN 0 240 51438 6

Library of Congress Cataloguing in Publication Data
A catalogue record for this book is available from the Library of Congress

Printed in Great Britain

Acknowledgements

Among the many people who helped make this book possible, I wish to express my gratitude for the enthusiasm and support of the following individuals:

Tom Davies for his enthusiasm for art and design education.
Jane Curry for her enthusiasm for photography in education.
Tim Daly for his advice and input to the final copy.
Bruce Gray for his unending technical support.
The students of Spelthorne College for their overwhelming enthusiasm and friendship, and finally to Dorothy, my partner, friend and wife.

Picture credits

Cover: Gareth Neal

Contents

Advanced study guides

Teaching Resources

Introduction to teachers

This book is intended as an introduction to photography for students studying GCSE, A level and GNVQ Art & Design courses. The emphasis has been placed upon a creative rather than technical approach to the subject.

A structured learning approach

The photographic study guides contained in this book offer a structured learning approach that will give students a framework for working on design projects and the photographic skills for personal communication.

The study guides are intended as an independent learning resource to help build design skills, including the ability to research, plan and execute work in a systematic manner. The students are encouraged to adopt a thematic approach, recording all developmental work in the form of background work or research sheets.

Flexibility and motivation

The study guides contain a degree of flexibility in giving students the choice of subject matter. This allows the student to pursue individual interests whilst still directing their work towards answering specific design criteria. This approach gives the student maximum opportunity to develop self-motivation. It is envisaged that staff will introduce each assignment and negotiate the suitability of subject matter with the students. Individual student progress should be monitored through group meetings and personal tutorials.

Implementation of the curriculum

The first three assignments are intended to be tackled sequentially and introduce no more technical information than is necessary for students to complete the work. This allows student confidence to grow quickly and enables less able students to complete all the tasks that have been set. The activities and assignments of the first three study guides provide the framework for the more complex assignments contained in the advanced section.

The activities and assignments contained in each study guide require approximately 25-40 hours contact time, plus additional student time devoted to independent research. For students studying at GCSE, GCE or GNVQ level each assignment would last approximately 6-8 weeks. For students studying photography full-time the complete curriculum contained in this book would cover the first two terms of the first year. The Media Studies guide 'Reading the Image' offers an optional photographic input for Media studies students or design students studying photography in greater depth.

Teaching resources

In the teaching resource section of this book you will find a work sheet and progress report which students should complete with the help of a member of staff. This process will enable the student to organise their own efforts and give valuable feedback to the student about their strengths and weaknesses. The controlled test should be viewed as another assignment which the student resources and then completes independently whilst being monitored. The students should be encouraged to demonstrate the skills which they have learnt in the preceding assignments.

Introduction to students

The study guides that you will be given on this course are designed to help you learn both the technical and creative aspects of photography. You will be asked to complete an assortment of tasks including research activities and practical assignments. The information and experience that you gain will provide you with a framework for your future design work with a camera.

What is design?

Something that has been designed is something that has been carefully planned. When you are set a photographic design assignment you are being asked to think carefully about what you want to take a photograph of, what techniques you will use to take this image and what you want to say about the subject you have chosen. Only when the photograph communicates the information that you intended, is the design said to be successful.

By completing all the activities and assignments in each of the study guides you will learn how other images were designed and how to communicate visually with your own camera. You will be given the freedom to choose the subject of your photographs. The images that you produce will be a means of expressing your ideas and recording your observations.

Design is a process which can be learnt as a series of steps. Once you apply these simple steps to new assignments you will learn how to be creative with your camera and produce effective designs.

Using the study guides

The study guides have been designed to offer you support during your design work. On the first page of each study guide is a list of aims and objectives laying out the skills covered and how they can be achieved. At the foot of each title page is a box titled 'notes'. This can be used to record important information such as assignment deadlines set by your tutor. The last page of each study guide is also left empty so that you can make notes during introductory talks about the assignment.

The activities are to be undertaken after you have first read and understood the supporting section on the same page. If at any time you feel unclear about what is being asked of you, consult a tutor.

Equipment needed

The course that you are following has been designed to teach you photography with the minimum amount of equipment. You will need a 35mm SLR camera with manual controls or manual override if automatic. Consult your teacher or a photographic specialist store if you are in doubt. Many dealers can supply second-hand equipment complete with a guarantee at reasonable prices. Large amounts of expensive equipment will not make you a better photographer. Many of the best professional photographers use less equipment than some amateurs. There are some areas of photography however which do require some very specialist equipment. These include some areas of sport and wildlife photography where you are unable to get very close to your subject. If these areas are of particular interest, you will need to think carefully about which of these sporting activities or animals are possible with the equipment you intend to use.

Research and resources

Introduction

The way to get the best out of each assignment is to use the activities contained in the study guides, as a starting point for your research.

You will only realise your full creative potential by looking at a variety of images from different sources. Creative artists and designers find inspiration for their work in different ways, but most find that they are influenced by other work they have seen and admired.

" The best designers are those who have access to the most information"

Stephen Bailey - former director of the Design Museum.

Getting started

Start by collecting and photocopying images that are relevant to the activity you have been asked to complete. This collection of images will act as a valuable resource for your future work. By taking different elements from these different images, e.g. the lighting technique from one and the vantage point from another, you are not copying somebody else's work but using them as inspiration for your own creation.

Talking through ideas with other students, friends, members of your family and with a teacher will help you to clarify your thinking, and develop your ideas further.

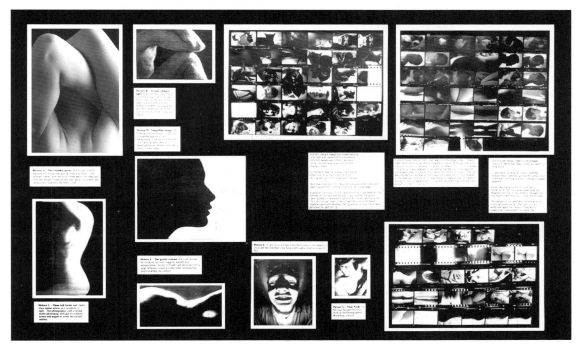

Student research sheet

Choosing resources

When you are looking for images that will help you with your research activities try to be very selective, using high quality sources. Not all photographs that are printed are necessarily well designed or appropriate to use. Good sources of photography may include high quality magazines and journals, photographic books and photography exhibitions. You may have to try many libraries to find appropriate material or borrow used magazines from friends and relatives. Keep an eye on the local press to find out what exhibitions are coming to your local galleries.

Presentation of research

In each assignment you are asked to provide evidence of how you have developed your ideas and perfected the techniques you have been using. This should be presented neatly and in an organised way so that somebody assessing your work can easily see the creative development of the finished piece of work.

You should edit your contact sheet including any alterations to the original framing with a chinagraph pencil or indelible marker pen. Make comments about these images to show how you have been selective and how this has influenced subsequent films that you have taken. You should clearly state what you were trying to do with each picture and comment on its success. You should also state clearly how any theme which is present in your work has developed.

Make brief comments about images that you have been looking at and how they have influenced your own work. Photocopy these images if possible and include them with your research.

All contact prints and photographs should be easily referenced to relevant comments using either numbers or letters as a means of identification. This coding will insure that the person assessing the work can quickly relate the text with the image that you are referring to.

Research and all contact prints should be carefully mounted on large sheets of card no larger than A1. Explanatory notes and comments should be made directly onto this sheet or on paper and glued into place.

Presentation of finished work

The way you present your work can influence your final mark. Design does not finish with the print. Try laying out the work on your card before sticking it down. Use rulers or a straight edge in aligning work if this is appropriate. Make sure the prints are neatly trimmed and make sure any writing has been spell-checked and is grammatically correct. Final work should be mounted on a sheet of card no larger than A1 using a suitable adhesive. Adhesives designed to stick paper do not work efficiently on resin coated photographic paper. Photographic prints are normally either dry mounted using adhesive tissue and a dry mounting press or window mounted. Both are time consuming and require a fair amount of skill. A cheaper and quicker alternative is to use a one centimetre strip of double sided tape on each corner of the photograph.

If you choose to write a title on the front of the sheet it is advisable that you either use a lettering stencil or generate the type using a computer. Be sure to write your name and project title on the back of this card so the person assessing the work can return the work to you quickly.

Storage of work

Assignment work should be kept clean and dry, preferably using a folder slightly larger than the size of your finished sheets. It is recommended that you standardise your presentation so that your final portfolio looks neat and presentable.

Negatives should always be stored in negative file sheets in dry dust free environments that will ensure clean reprints can be made if necessary.

Technical Section

THE CAMERA

The Camera - Mark Galer

Aims

The aims of this guide are:
~ To offer an independent resource of technical information.
~ To develop knowledge and understanding about camera equipment.

Objectives

These are the things you will be expected to do:
~ Operate the basic controls of a 35mm SLR camera.
~ Adjust shutter speed and aperture in response to taking light meter readings.
~ Correctly expose a roll of film.
~ Care for camera equipment.

Notes

Introduction

At first glance the 35mm SLR camera is a sophisticated and confusing piece of equipment. Just as with any other complex piece of machinery the user can, over a period of time, become very familiar with its operations and functions until they are almost second nature. Operating a 35mm SLR camera, just like driving a car, is a skill which is attainable by most people. How long it will take to acquire this skill will very much depend on the individual and how much time and effort they are prepared to spend operating the camera. Different makes and models of 35mm SLR cameras may appear very different but they all share the same basic features, its just that they may be placed in different locations on the camera body or operated automatically. If you have any difficulty in finding a particular feature on your own camera ask a teacher or consult your camera manual to help you locate it.

Why a 35mm SLR?

The 35mm SLR is the most popular camera used by the keen amateur and professional photographer. The camera is both flexible due to the large range of accessories available and light enough carry around. The 35mm SLR uses 35mm film which comes in ready loaded cassettes that are very quick to load. Single Lens Reflex is a term which is used to describe the way we view the image through this type of camera. The SLR camera has a single lens unlike compact cameras which have two. The compact camera has one lens for viewing the subject and another for capturing the image onto the film. The single lens of the SLR camera is used to view the subject and take the picture. This is achieved by the use of a mirror behind the lens which reflects the image up into the viewfinder via the Pentaprism which converts the mirror image to one which appears the right way round. If we change the lens, use a coloured filter or change the focus we can see all of these changes through the viewfinder.

Automatic or manual

If your camera operates automatically spend some time finding out how the camera can be switched to manual operation or how you can override the automatic function. Automatic cameras are programmed to make decisions which are not necessarily correct in every situation. A good photographer must be able to use the manual controls of the camera.

Care of the camera

To take good care of your camera you need to know several important points.

1. Avoid dropping your camera - use a strap to secure it around your neck or wrist.
2. Avoid getting your camera wet - when it is raining place the camera under your jacket or in a camera bag.
3. Only clean your camera lens with a soft brush or a lens cleaning cloth.
4. Never touch the mirror inside the mirror housing or the shutter curtain inside the film back. Both these items are extremely delicate.

NB. Damage to your camera can usually only be repaired by a camera specialist and will usually incur a minimum fee which can be greater than the value of your camera.

THE 35mm SLR CAMERA

The basic controls

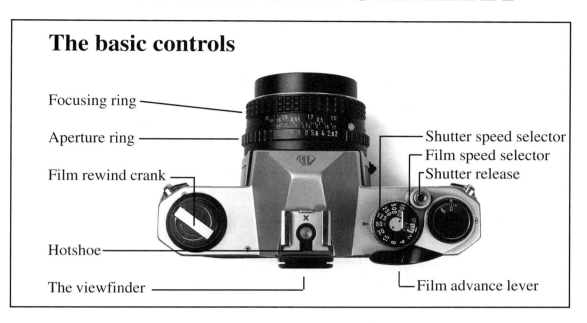

Focusing ring

Aperture ring

Film rewind crank

Hotshoe

The viewfinder

Shutter speed selector
Film speed selector
Shutter release

Film advance lever

Film back

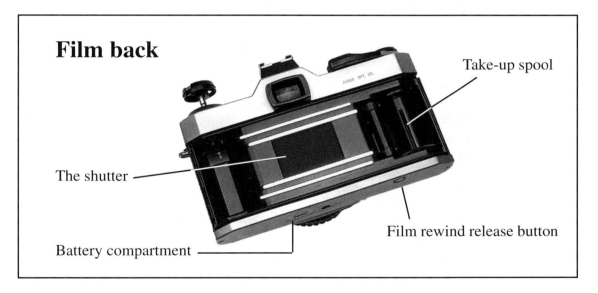

Take-up spool

The shutter

Battery compartment

Film rewind release button

Mirror housing

Pentaprism

Lens
alignment mark

Lens release

Mirror

X sync socket

THE FIRST FILM

Open the back of the camera. This is usually done by pulling the rewind crank up.

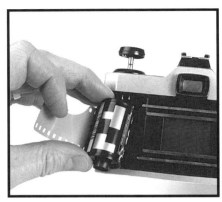

Load the film cassette. The rewind crank must be lifted to enable the cassette to be inserted.

Push the rewind crank down and pull a short length of film out of the cassette.

Attach the film leader to the take-up spool. Most of the film leader can be inserted.

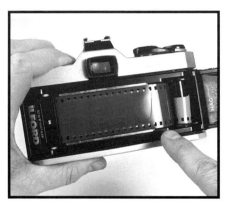

Fire the shutter, advance the film. The teeth must engage with the film sprocket holes.

Take up any slack in the film by rewinding the crank handle gently. Close the back securely.

Fire two blank shots and advance the film. Check the rewind crank turns each time you advance the film to ensure the film is advancing.

Set the film speed or ISO that appeared on the film cassette or box on the film speed selector. This is not moved again whilst exposing this roll of film.

Select a shutter speed. Usually nothing slower than 1/60th of a second unless you are using a tripod.

THE FIRST FILM

Move the focusing ring back and forth until your subject appears as sharp as possible.

Adjust the aperture until you obtain the correct exposure. See page

Hold the camera firmly, frame your shot and press the shutter release gently.

Advance the film using the film advance lever. The shutter will only fire when fully advanced.

When the last frame has been taken push the film rewind button.

Rewind the film smoothly back into the cassette, approximately one complete turn per frame.

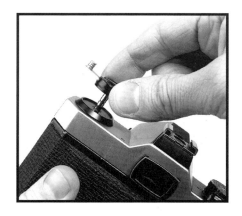

When you feel the film tighten and then go slack or hear the noise of the film coming away from the take-up spool the film has rewound.

Remove the film from the camera by pulling the rewind crank up. If the film leader is still visible, mark the film to indicate that it is exposed.

Place the film back into its container until you are ready to process it. This will protect the film from moisture and dirt.

Exposure

The aperture

The aperture of the camera lens opens and closes like the iris of the human eye. Just as the human iris opens up in dim light and closes down in bright light to control the amount of light reaching the retina, the aperture of the camera lens must also be opened and closed in different lighting conditions to control the amount of light that reaches the film. The film requires exactly the right amount of light to create an image. Too much light and the film will be overexposed and appear very dark or dense once it is processed. Not enough light and the film will be underexposed and appear very light or 'thin'.

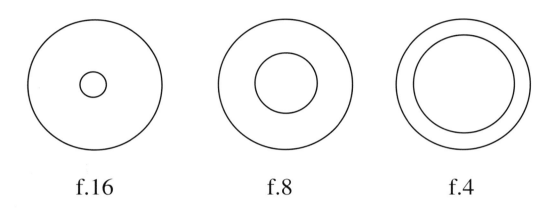

f.16   f.8   f.4

We are guided to select the correct aperture by the light meter. As the aperture ring is turned on the camera lens a series of clicks can be felt. These clicks are called f-stops and are numbered. Depending upon which way you turn the aperture ring every stop lets exactly twice or half as much light reach the film as the previous one. The light meter will advise you when you are getting closer to the correct aperture. The highest f-number corresponds to the smallest aperture and the smallest f-number corresponds to the widest aperture.

The shutter

When the shutter button is pressed the shutter opens for the amount of time set on the shutter speed dial. These figures are in fractions of seconds. The length of time the shutter is open also controls the amount of light that reaches the film, each shutter speed doubling or halving the amount of light. Exposure, therefore, is a combination of aperture and shutter speed.

To slow the shutter speed down is to leave the shutter open for a greater length of time. Shutter speeds slower than 1/60th of a second can cause movement blur or camera shake unless you hold the camera steady with a tripod or by some other means.

To use a shutter speed faster than 1/250th requires a wide aperture or a very light sensitive film to compensate for the small amount of light that can pass through a shutter that is open for such a short amount of time. It is suggested that you keep the shutter speed dial on 1/60th or 1/125th of a second until you are familiar with the equipment that you are using.

Using the light meter

Most light meters in 35mm SLR cameras will show the light meter reading inside the viewfinder. This allows the photographer to alter the camera settings to achieve the correct exposure without having to remove the camera from the eye.

Inside the viewfinder you should see either a needle, a series of lights or an LED display which either move or change when the camera is pointed towards the light. The light meter in a 35mm SLR camera requires a healthy battery to work so if you cannot see any changes consult a teacher or camera dealer.

Most 35mm SLR cameras will have a metering system which will resemble one of the four below. If your metering system is different seek advice.

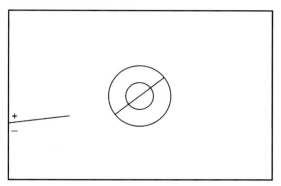

1. Position the needle between the + and - symbols by altering the aperture or shutter speed dial. The box containing the + and - symbol may be replaced by a second needle or a circle.

2. One light indicates the shutter speed that has been selected. Move the other light or lights towards this one by altering the aperture until only one light shows.

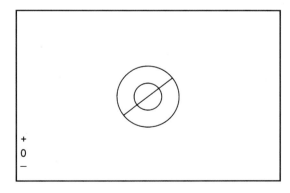

3. Move the aperture or shutter speed until the symbol '0' is illuminated.
Too much light is indicated by a +
Too little light is indicated by a -.

4. Decrease or increase exposure as guided by a series of bars which are displayed if there is insufficient or excessive exposure either side of the 0 symbol.

Accurate metering

It is important when taking a light meter reading to avoid pointing the centre circle of the viewfinder towards a bright light or something very dark. Try to position the centre circle on the subject that you are taking to find the correct exposure. Set the camera and then frame your shot.

Notes

PROCESS & PRINT

Mark Galer

Aims

The aims of this guide are:
~ To offer an independent resource of technical information.
~ To develop knowledge and understanding about the technical processes and
procedures involved with producing photographic images.

Objectives

These are the things you will be expected to do:
~ Develop black & white photographic film in daylight developing tanks.
~ Produce contact sheets from black & white film for editing purposes.
~ Produce clean black & white prints with acceptable sharpness and contrast.

Notes

Film Processing

- You must use a light-tight darkroom.
- Lock the door behind you if possible.
- You must have with you a <u>DRY</u> film spool, black centre column for the spool, developing tank, tank lid, scissors and film cassette opener if needed.

With the Lights On
If you have rewound your film leaving the leader out, or have managed to retrieve the leader using a special device, the following can be done with the lights on.
- Cut the film leader off between the sprocket holes (leave no sharp edges)
- Feed the first four centimetres of film onto the spool.

With the Lights Off
If the film leader has been wound back into the cassette the following will have to be carried out in total darkness. It is recommended that you practice loading a blank film onto a spiral several times before attempting this in the dark.
- Open the cassette in total darkness with a special tool similar to a bottle opener.
- Gently withdraw the film from the cassette and remove the film leader with a pair of scissors. Check that no sharp edges are left.
- Let the film hang down and push the film gently onto the spiral.

Loading the film in total darkness
- Wind the film gently onto the spiral by twisting one side of the spiral back and forth.
- If the film jams, remove it from the spiral and start again. Check the spool is dry and the leading edge of the film is not damaged.
- If the film breaks or tears, place the film in a dry tin or the developing tank which has been sealed with the waterproof lid and seek assistance.
- When the film is wound onto the spiral cut off the film spool.
- Push the loaded spiral onto the centre column and then place into the tank. Check that centre column base is placed down towards the base of the tank.
- Place the lid on the tank ensuring that the lid is not twisted in the process.
- Check with the other students before switching any lights on.

Preparing the chemicals
- Fill a 2 litre jug with water at 20°C. Use a thermometer and mix hot and cold water if necessary. If the film developer is very cold or warm this temperature may be raised or lowered to compensate.
- Measure accurately with a measuring cylinder the correct amount of film developer for each film to be developed (If the developer is concentrated a small measuring cylinder should be used to gain an accurate reading).
- Pour the developer into a measuring cylinder containing enough water to make up the total quantity needed - 300 ml for one film, 600 ml for two, 900 ml for three.
- Have ready measuring cylinders with the required amount of stop and fix to complete the process.

FILM PROCESSING 1

Lay out the equipment so that you will be able to find everything in the dark.

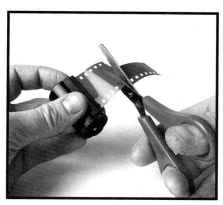

Cut off the leader between the sprocket holes and feed a few centimetres onto the spiral.

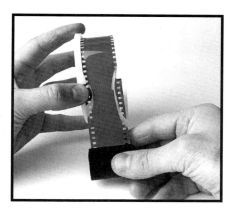

Switch off the light and pull the film cassette gently down until about a metre of film is free.

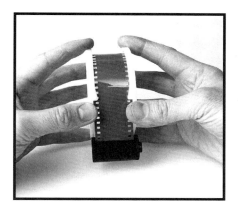

Place your thumbs behind the entrance to the gate as you wind the film onto the spiral.

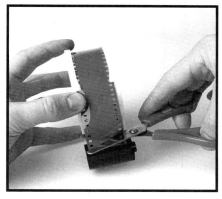

When no more film is left in the cassette, remove it using a pair of scissors.

Place the spiral onto the column and ensure the lip is facing towards the base of the tank.

Turn the lid of the tank clockwise until it locks into place. The lid should turn freely until it locks, otherwise it may be out of alignment.

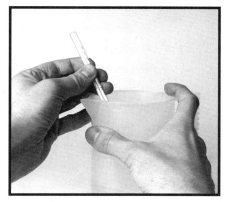

Prepare a jug full of water at 20°C. Measure this accurately with a thermometer and adjust the temperature with hot or cold water if necessary.

Measure out the film developer using a small measuring cylinder and make up to the required amount using the water from the jug.

17

Developing

- Pour the solution into the tank quickly and start timing immediately.
- Press on waterproof tank cover firmly and invert the tank several times.
- Every 30 seconds give 2 inversions of the tank.
- Start pouring developer away 15 seconds before the correct time has elapsed. Developer is not to be used again.

Stop

- Fill the tank without delay with STOP BATH from the measuring cylinder.
- Replace the cover and invert for 30 seconds.
- Return the stop bath to the correct container using a funnel.
- If the stop bath turns blue, pour it away.

Fixing the film

- Fill the tank with fixer from the measuring cylinder which has already been diluted.
- Replace the cover and invert once every 15 seconds for 5 minutes.
- Return the fixer to the correct container using a funnel.

Washing the film

- Fill the tank with water from the cold tap.
- Replace the cover and invert the tank several times.
- Pour the water away and remove the tank lid.
- Check that the film is clear (no milkiness)
- Wash under running water for 15-20 minutes.

Final rinse

- Remove the films from the tank and fill it with clean water.
- Add a few drops of wetting agent to the water.
- Replace the films and waterproof seal, invert twice and leave for 30 seconds.

Drying the film

- Remove the films from the tank and remove from the film spiral.
- If you are in a hurry wipe the excess water off the films. This can be done by wiping the film between two clean fingers. Dirty or damaged squeegees will scratch or 'tram-line' your film.
- Hang the films up to dry with a weight on the bottom so that they dry straight. Do not hang the films to dry in a dusty place or near a radiator.
- Do not dry the film with a hairdryer as this may damage the film.

Clearing away

- Replace all the lids on the chemical bottles.
- Rinse all parts of the tank and mixing equipment and leave to drain.
- Clean and tidy away all materials.
- Film should dry in 10 minutes in the film dryer, 2 hours if hung on a line.
- If using heat to dry your film do not leave it too long or have the temperature too high. Both of these actions will lower the quality of your film.

FILM PROCESSING 2

Pour in the developer quickly, start timing and attach the liquid seal to the tank.

Invert the tank several times and then twice every 30 seconds.

After the correct time has elapsed pour the developer away.

Fill the tank with stop bath, replace the cover and invert for 15 seconds.

Pour the stop bath back into its container and then fill the tank with fixer.

Save the fix, rinse the tank out with water and then inspect the film to ensure that it is clear.

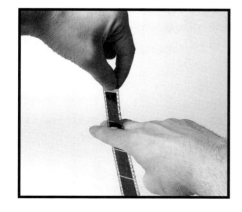

Wash the film for 20 minutes. Add a few drops of wetting agent to the final wash. Shake all water from the spiral or wipe the film between 2 fingers.

Hang the film up to dry in a dust free area with a weight attached to the base. Wash all the equipment and surfaces that have been used.

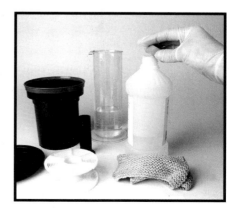

Replace all of the tops on the chemical containers and put the tank and spiral in a warm place to dry.

Wet processing

Introduction

Producing your own prints is a rewarding photographic activity. To get the best results and avoid wasting materials you should adopt a methodical and patient approach.

Health & safety

It is important to protect yourself and your clothes from chemical contamination when processing photographic paper. Although the chemicals are not highly corrosive they can lead to rashes and stained clothes. The following precautions must be taken.

a) Wear a laboratory coat or old shirt at all times whilst processing.
b) When pouring chemicals wear protective gloves and eye protection.
c) Never handle prints with your hands. Use the print tongs provided.
d) Wash contaminated skin or clothes immediately with soap and water.
e) Never sit on work surfaces that have come into contact with chemicals.
f) Never bring food or drink into an area where chemicals are being used.
g) Ensure that the room you are working in is well ventilated.

Processing procedure for resin coated material

1. Place the exposed piece of photographic paper into a tray of developer at room temperature (approximately 20°C) and submerge it using a pair of print tongs.
2. Rock the tray gently for 1 minute. Do not 'snatch' the paper from the developer if you think it is going too dark. This will only lead to a poor quality print.
3. Use print tongs to transfer the paper to the tray containing stop bath (allow a few seconds for the surplus developer to drip back into the developer tray). The tongs from the developer tray must not touch the stop bath. If they do, you must wash them in water before returning them to the developer tray.
4. After 10 seconds transfer the paper to the fix using print tongs from the fix tray.
5. After 2 minutes in the fix transfer the print to running water for a further 2 minutes. If another print from the fix is added to the water during the wash period the original print must be washed for another 2 minutes.
Test strips may be examined in daylight conditions after only 15 seconds in the fix and 5 seconds in the wash if they are to be thrown away after printing.
6. Squeegee the water off both surfaces of the print applying light pressure only to avoid scratching the print.
7. Dry the prints face-up on a clean and dry work surface in a warm room. Prints must not overlap during the drying period as they will stick together. Inspect the surface of the print after 10 minutes for droplets of water before stacking the prints. Any droplets of water sitting on the print surface can be removed by carefully dabbing them with a dry paper towel and then leaving them for an additional 5 minutes. Drying prints with excessive heat will cause the prints to curl badly.

Cleanliness

Dirty marks hairs or white spots appear due to chemical contamination or dirty negatives ... Make sure your hands and work bench are clean and dry. Use a blower brush to clean negatives. **HANDLE NEGATIVES WITH CARE.**

Test strips

Test strips are used to measure the brightness of the light coming from the enlarger lens. They are used for the same reason that we take a light meter reading in the camera, i.e. to find the correct aperture and the length of time needed for a correct exposure. Because photographic paper requires more exposure than film we use a timer connected to the enlarger to measure seconds, instead of the camera shutter which measures fractions of seconds. It is however possible to count accurately or use the second hand of a clock to measure exposures for photographic paper.

There is the temptation to guess an exposure from the first test strip even if all the exposures shown on the test strip are far from correct. If used often, test strips will save you time and photographic paper.

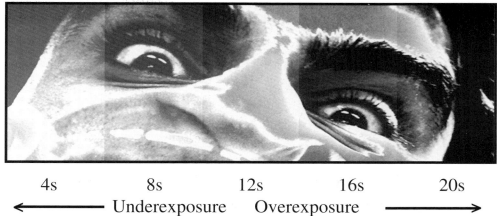

| 4s | 8s | 12s | 16s | 20s |

←———————— Underexposure   Overexposure ————————→

A good test strip will cross many areas of density and show over and under exposure.

Contrast control

The term contrast refers to the different tones of grey we can see in a print
A low contrast print is one that may have many shades of grey but no deep blacks and no clean whites. A high contrast print is one that has many areas of black and white but few, if any, shades of grey. The contrast can be altered on multigrade paper by changing the colour of the filter in the enlarger head.

Low contrast   *High contrast*

Increasing the Magenta filter on a colour enlarger will increase the contrast on multigrade paper. Increasing the Yellow filter will lower the contrast. Increasing the filtration on an enlarger will require longer exposures to obtain a print of the same density.

Increasing to a higher number on a multigrade head will increase the contrast of the final print. Decreasing the number on the multigrade head will decrease the contrast of the final print.

THE HIGHER THE NUMBER THE GREATER THE CONTRAST.

Making a contact sheet

Preparing your negatives

1. Cut your negatives into strips of five or six frames on a clean dry surface. Use sharp scissors and be careful not to cut into any of the images.
   Do not handle the image area of the negatives with your fingers.
2. Slide the negative strips into the negative file shiny side up (emulsion side down). Store your negatives away from wet or damp work surfaces.
3. Choose an enlarger and switch off any white lights.
4. Switch on the enlarger and adjust the lens aperture so that the light is at its brightest. Close the aperture by 2 stops.
5. Adjust the enlarger head height until the spread of light covers the negative bag or contact printing frame. Switch off the enlarger.

Making a test strip

7. In red light only, cut a 2cm strip off a sheet of photographic paper using a pair of scissors or a guillotine. Return the rest of the paper to the bag and seal it.
8. Place the strip of photographic paper emulsion side up on the foam of the contact printing frame or on the baseboard if you have a separate piece of glass (on your pearl or gloss paper the emulsion side will reflect the red lights).
9. Place an average strip of negatives, neither the darkest nor the lightest, on top of the strip of paper. If you have a clear negative file you do not need to remove them.
10. Cover the photographic paper and the negative with the clean sheet of glass and ensure that both are in close contact. Press down if necessary.
11. Cover up 3/4 of the test strip with a piece of card.
12. Set the enlarger timer to 4 seconds and expose the first 1/4 of the strip.
13. Uncover one further 1/4 and expose for a further 3 seconds. Repeat process until all the paper has been exposed.
14. Remove the photographic paper from beneath the glass.
15. Process the test strip in trays of developer, stop and fix.
16. Examine the test strip in daylight conditions. Choose the best exposure from the test strip and set the enlarger timer accordingly. If all the exposures are too light then increase the exposure times for a second test strip. If all the exposures are too dark close the lens aperture down 1 more stop.
17. Expose all of the negative strips onto one sheet of printing paper and process.

REMEMBER: THE BRIGHTER THE LIGHT, THE DARKER THE PRINT.

Common faults

1. Faint and fuzzy image .... Photographic paper upside down.
2. Test strip too dark .... Enlarger lens not stopped down.
3. Test strip too light .... Enlarger lens stopped down too much.
4. Out of focus images .... Glass not in close contact.
5. Marks or splodges .... Glass not clean or chemical contamination.

MAKING A CONTACT SHEET

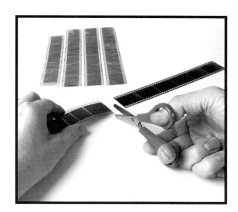

Cutting negatives. Be careful not to drag negatives along the work surface.

Sliding negatives into file sheet. Handle negatives carefully by the edges to ensure negatives stay clean.

Stopping lens down 2 stops. This will ensure that the exposure time is of a reasonable length.

The spread of light from the enlarger lens must cover the baseboard.

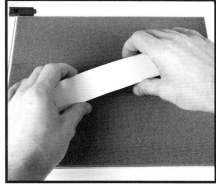

Placing the test strip, emulsion side up, on the baseboard of the enlarger.

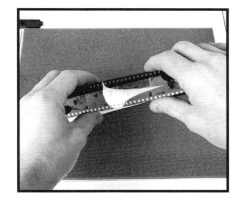

Placing the negative onto the test strip, emulsion side down, shiny side up.

Placing a clean sheet of glass over the negative and photographic paper.

Masking off portion of test strip using a piece of thick card to control exposure.

Processing photographic paper.

Printing

1. Select a negative to print from the contact sheet. Ensure the enlarger is switched off at the timer or the power source.
2. Remove the negative carrier from the enlarger.
3. Place the negative in the carrier, emulsion side down, glossy side up.
4. Blow any dust off the negative using a blower brush.
5. Replace the negative carrier in the enlarger.
6. Place printing easel on the enlarger base board and switch on the enlarger.
7. Taking care not to touch lens surface, make sure the aperture is at its largest f-stop to supply maximum light for focusing.
8. Raise or lower the enlarger head to obtain the desired degree of image enlargement and reposition the easel.
9. Focus the image (use focusing aid for precise focusing).
10. Compose and frame the picture.
11. Reduce the aperture of the enlarger lens by 2 or 3 f-stops. The smaller opening offsets slight focusing errors.
12. Turn the enlarger off and place a test strip on the easel, emulsion side up. **See section on test strips for more information.**
13. Expose the test strip at 4 second intervals.
14. Process test strip.
15. Select the best exposure in daylight.

If the exposure time is less than 7 seconds, close the lens down 1 more f-stop. If it is greater than 30 seconds open up the lens 1 f-stop.
EACH F-STOP WILL DOUBLE OR HALVE THE EXPOSURE TIME.

Burning in

When a section of the print appears too light it is possible to increase the exposure to this area without affecting the rest of the print. This is called 'burning in'.
To burn in use a large piece of card with a hole cut in it. The shape of the hole can vary depending upon the area to be burnt in, but it should be much smaller than the area needing the extra exposure. Using test strips, find out the additional exposure needed for the area to be burnt in. Expose the entire paper for the shorter exposure and then place the card with the hole approximately midway between the lens and the paper for the additional exposure. The light falling through the hole should be directed onto the area needing the additional exposure and the card should be kept moving during this exposure so that no hard edge between the original and extra exposure appears on the final print.

Holding back or dodging

When a small area of the print appears too dark it is possible to reduce the exposure to this area without effecting the rest of the print. This is called 'dodging'.
The selected area is shaded for part of the exposure by placing a small piece of card taped to a piece of wire between the enlarger lens and photographic paper. The amount of time the area needs to be shaded can be determined using a test strip, and as in the burning in process the card should be placed midway between the enlarger lens and the photographic paper and kept moving. The wire should be rotated slowly like the second hand of a clock so that a white line does not appear in the final image.

PRINTING

Place the negative into the carrier, shiny side up, emulsion side down.

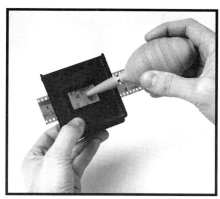

Remove any dust from the negative using a blower brush or soft cloth.

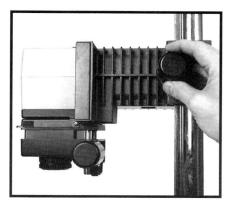

Adjust the height of the enlarger head to create an image of approximately the right size on your printing easel.

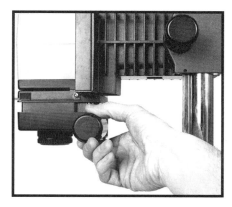

Focus the image by eye with the enlarger lens set to its maximum aperture.

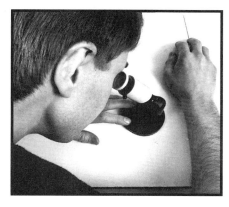

Use a focus finder to ensure accurate focusing.

Reduce the aperture by 2 stops to ensure a reasonable length of exposure and sharp focus.

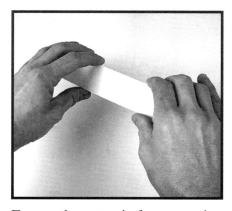

Expose the test strip from a section of the projected image which has a variety of densities.

Process the test strip. Ensure that the test strip is processed for the full processing time.

Viewing the test strip to gauge the correct exposure and contrast before making the final print.

Notes

Notes

Study Guides

THE FRAME

Pavement - Gareth Neal 1991

Aims

Your aims are:
~ To develop an awareness of how a photographic print is a two-dimensional composition of lines, shapes and patterns.
~ To develop an understanding of how different ways of framing can affect both the emphasis and the meaning of the subject matter.

Objectives

These are the things you have to do:
~ **Research** - Produce a study sheet looking at the composition of different photographs and the techniques employed by the photographers.
~ **Discussion** - Exchange ideas and opinions with other students.
~ **Practical Work** - Produce photographic prints through close observation and selection that demonstrate how the frame can create compositions of shape, line and pattern and a personal theme.

Notes

Introduction

From photographs each of us can learn more about the world. Images not only inform us about the products we never knew we needed, the events, people and places too distant or remote for us to see with our own eyes, but also tell us more about the things we thought we already knew.

Most of us are too preoccupied to stand and look at something for any great amount of time. We glance at something briefly and think we have seen it. Our conditioning or desires often tell us what we have seen or would like to see. When we look at a photograph of something ordinary however it may show us the object as we had never seen it before. With a little creative imagination and a little photographic technique it is possible to release the extraordinary from the ordinary.

Bill Brandt in 1948 said that "it is the photographer's job to see more intensely than most people do. He must keep in him something of the child who looks at the world for the first time or of the traveller who enters a strange country."

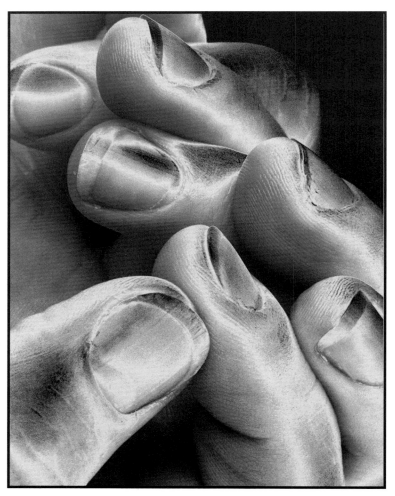

Fingers - Mark Galer

Choosing a subject

In order to photograph something that will be of interest to others you must first remove the blinkers and photograph something that is of interest to you. Your first creative decision is an important one. What will you choose to photograph? Your first technical decision is how to frame it.

Composition

Framing the Subject

The photographer Robert Capa said "If your pictures aren't good, you're not close enough". He was saying that the subject matter can look unimportant and not worthy of closer attention. There is also a danger that the photographer will not have control over the composition.

A common mistake made by many amateur photographers is that they stand too far away from their subject matter, in a desire to include everything. Their photographs become busy, unstructured and cluttered with unwanted detail which distracts from the primary subject matter.

Embrace - Eikoh Hosoe 1970

The photograph above is a study of the human figure and also a composition of shape, tone and line. There are three dominant shapes. The woman's leg, the man's back and also the third shape which is created between the frame and the man's back. The act of framing a subject using the viewfinder of the camera imposes an edge that does not exist in reality. This frame also dissects familiar objects to create new shapes. The shapes that this frame creates must be studied carefully in order to create successful compositions.

The powerful arc of the man's back is positioned carefully in relationship to the edge of the frame and the leg of the woman is added to balance the composition.

Filling the frame

When the photographer moves closer, distracting background can be reduced or eliminated. There are less visual elements that have to be arranged and the photographer has much more control over the composition. Many amateurs are afraid of chopping off the top of someone's head or missing out some detail that they feel is important. Unless the photograph is to act as a factual record the need to include everything is unnecessary.

Father and Child, Vietnam - Marc Riboud

When the viewer is shown a photograph they have no way of knowing for sure what lies beyond the frame. We often make decisions on what the photograph is about from the information we can see. We often have no way of knowing whether these assumptions are correct or incorrect.

The photograph above is of a father and child. The protective hands of a father figure provide the only information most people need to arrive at this conclusion. In order to clarify any doubt the photographer may have decided to move further back to include the whole figure. The disadvantage in doing this would have been that the background would also begin to play a large part in the composition and the power of this portrait of a child and his father would have been lost. Photographers do have the option however of taking more than one photograph to tell a story.

Activity 1

Look through assorted photographic books and observe how many photographers have moved in very close to their subjects. By employing this technique the photographer is said to 'fill the frame' and make their photographs more dramatic.

Find two examples of how photographers seek simple backgrounds to remove unwanted detail and to help keep the emphasis or 'focal point' on the subject.

The whole truth?

Photographs provide us with factual information but sometimes we do not have enough information to be sure what the photograph is about.

Hyde Park - Henri Cartier-Bresson

What do we know about this old lady or her life other than what we can see in the photograph? Can we assume she is lonely as nobody else appears within the frame? Could the photographer have excluded her grandchildren playing close at hand to improve the composition or alter the meaning? Because we are unable to see the event or subject that the photograph originated from we are seeing it out of context.

Activity 2

Read the following passage taken from the book *The Photographer's Eye* by John Szarkowski and answer the questions below.

"To quote out of context is the essence of the photographer's craft. His central problem is a simple one: what shall he include, what shall he reject? The line of decision between in and out is the picture's edge. While the draughtsman starts with the middle of the sheet, the photographer starts with the frame.
 The photograph's edge defines content. It isolates unexpected juxtapositions. By surrounding two facts, it creates a relationship. The edge of the photograph dissects familiar forms, and shows their unfamiliar fragment. It creates the shapes that surround objects.
The photographer edits the meanings and the patterns of the world through an imaginary frame. This frame is the beginning of his picture's geometry. It is to the photograph as the cushion is to the billiard table."

Q. What does John Szarkowski mean when he says that photographers are quoting 'out of context' when they make photographic pictures?

Q. The frame often 'dissects familiar forms'. At the end of the last century photography was having a major impact on Art. Impressionist artists such as Degas were influenced by what they saw. Find an example of his work which clearly shows this influence and explain why the public might have been shocked to see such paintings.

Designing the image

When the photographer has chosen a subject to photograph there is often the temptation, especially for the untrained eye, to place the subject in the middle of the picture without considering the overall arrangement of shapes within the frame. If the focal point is placed in the centre of the frame, the viewers eye may not move around the whole image and this often leads to a static and uninteresting composition. The design photographer should think carefully where the main subject is placed within the image, only choosing the central location after much consideration.

"A picture is well composed if its constituents - whether figures or apples or just shapes - form a harmony which pleases the eye when regarded as two-dimensional shapes on a flat ground." (Peter and Linda Murray - A Dictionary of Art and Artists).

The rule of thirds

Rules of composition have been formulated to aid designers create harmonious images which are pleasing to the eye. The most common of these rules are the 'golden section' and the 'rule of thirds'.

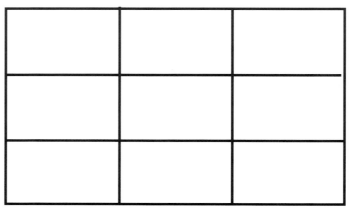

The rule of thirds

The Golden Section is the name given to a traditional system of dividing the frame into unequal parts which dates back to the time of Ancient Greece. The rule of thirds is the simplified modern equivalent. Try to visualise the viewfinder as having a grid which divides the frame into three equal segments, both vertically and horizontally. Many photographers and artists use these lines and their intersection points as key positions to place significant elements within the picture.

Breaking the Rules

Designers who are aware of these rules often break them by deliberately placing the elements of the image closer to the edges of the frame. This can often be effective in creating dynamic tension where a more formal design is not needed.

Activity 3

Find two examples of photographs that follow the rule of thirds and two examples that do not. Comment briefly on why and how you think the composition works.

Use of line

The most powerful design element in any black and white photograph is the positioning of the dominant line or lines. This could be a line that is well defined such as the horizon line in a landscape photograph or it can be a 'suggested' line that appears to flow through the linked arms of several people.

Diagonal lines

Whether real or suggested, these lines are more dynamic than horizontal or vertical lines. Whereas horizontal and vertical lines are stable, diagonal lines are seen as unstable (as if they are falling over) thus setting up a dynamic tension or sense of movement within the picture.

Columns - Mark Galer

A dominant line that curves

A dominant line that is present in the subject matter is often arranged in the viewfinder in the shape of a spiral or S-curve and is very useful in drawing the viewer's eye through the picture in an orderly way. The viewer often starts at the top left hand corner of the image and many S-curves exploit this. Curves can be visually dynamic when the arc of the curve comes close to the edge of the frame.

Activity 4

Find two examples of photographs that use straight lines as an important feature in constructing the pictures' composition. Find one example where the dominant line is either an arc or S-curve. Comment briefly on the contribution of line to the composition of each example.

Vantage point

A carefully chosen viewpoint or 'vantage point' can often reveal the subject as familiar and yet strange. In designing an effective photograph that will encourage the viewer to look more closely, and for longer, it is important to study your subject matter from all angles. The 'usual' or ordinary is often disregarded as having been 'seen before' so it is sometimes important to look for a fresh angle on a subject that will tell the viewer something new.

The Beach - Joanne Arnold

When we move further away from our subject matter we can start to introduce unwanted details into the frame that begin to detract from the main subject. Eventually the frame is so cluttered that it can look unstructured. The careful use of vantage point can sometimes overcome this. A high or low vantage point will sometimes enable the photographer to remove unwanted subject matter using the ground or the sky as an empty backdrop.

Activity 5

Find two examples of photographs where the photographer has used a different vantage point to improve the composition. Comment on how this was achieved and how this has possibly improved the composition.

Depth

When we view a flat two-dimensional print which is a representation of a three-dimensional scene, we can often recreate this sense of depth in our mind's eye. Using any perspective present in the image and the scale of known objects we view the image as if it exists in layers at differing distances. Successful compositions often make use of this sense of depth by strategically placing points of interest in the foreground, the middle distance and the distance. Our eye can be led through such a composition as if we were walking through the photograph observing the points of interest on the way.

Doctor's Waiting Room, Battersea, London 1975 - Ian Berry

In the image above our eyes are first drawn to the largest figures occupying the foreground on either side of the central doorway. In a desire to learn more from the image our eyes quickly progress towards the figures occupying the middle distance. Appearing as lazy sentinels the figures lean against the doorway and move our gaze towards the focal point of the photograph, the small girl holding her mother's hand in the centre of the image. The technique of drawing us into the photograph is used in many photographs and can be also be exploited using dark foreground tones drawing us towards lighter distant tones.

Activity 6

Find two photographs where the photographer has placed subject matter in the foreground, the middle distance and the distance in an attempt either to fill the frame or draw our gaze into the image. Comment briefly on where you feel the focal points of these images are.

Practical assignment

Produce a set of six photographs investigating natural or man-made forms. Your work should demonstrate how the frame can be used to create compositions of shape, line and pattern. You should consider not only the shapes of your subject matter but also those formed between the subject and the frame.

Choosing a theme

Your photographs should develop a clearly defined theme. This could be several different ways of looking at one subject or different subjects that share something in common, e.g. a similar pattern or composition. If you are unfamiliar with your camera choose a subject that will keep still, allowing you time to design the composition.

A possible title for your set of prints could be:

1. Patterns in nature
2. Rhythms of life
3. Urban patterns

Your work should:

a) make use of differences in subject distance including some work at, or near, the closest focusing distance of your camera lens.

b) show that you have considered the rule of thirds.

c) demonstrate the creative use of line in developing your compositions.

d) make use of different vantage points.

e) show that you have thought carefully about the background and the foreground in making your composition.

NB. Remember to implement aspects of your research during your practical assignment.

Resources

1. The Image - Collins Photography Workshop .... Michael Freeman

2. Photographic Composition ....Tom Grill & Mark Scanlon

3. Basic Photography - Pages 142-153 .... Michael Langford

4. The Great Themes - Pages 52-59 .... Time-Life on Photography

Spiral Staircase - Tom Scicluna 1991

Many photographers use the lines and intersection points of the rule of thirds as key positions to place significant elements in the picture.

Arc - Tom Scicluna 1991

By tilting the camera the shapes and lines can be carefully organised within the frame. The dominant arc in this image sweeps the edge of the frame and exits at the bottom right hand corner.

Staircase - Philip Leonard 1992

Diagonal lines that appear in a picture, whether real or suggested, are more dynamic than horizontal lines. The diagonal lines have been arranged to enter the corners of this photograph.

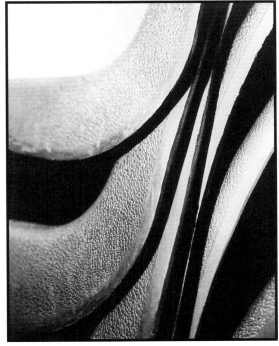

Stacked Chairs - Gareth Neal 1991

A carefully chosen viewpoint can often reveal the subject as familiar and yet strange. A student has explored the interesting lines created by a stack of chairs.

Notes

LIGHT

Hands - Ashley Dagg-Heston

Aims

Your aims are:
~ To develop knowledge and understanding of how the directional use of light can change character and mood.

Objectives

These are the things you have to do:
~ **Research** - Produce a study sheet that looks at the 'atmospheric' lighting of several different photographs and the techniques employed by the photographers to achieve it.
~ **Discussion** - Exchange ideas and opinions with other students on the work you are studying.
~ **Practical Work** - Produce photographic prints through close observation and selection that demonstrate how the type of light and its direction affects the way we view the subject matter.

Notes

Introduction

Light is the black and white photographer's medium. The word photography is derived from the ancient Greek words photos and graph meaning 'light writing'. In photography, shadows, reflections, patterns of light, even the light source itself may become the main subject and solid objects may become incidental to the theme.

Projection - Student photograph

Creating atmosphere

Directional light from the side and/or from behind a subject can produce some of the most evocative and atmospheric black and white photographs. Most snapshots by amateurs however are taken either outside when the sun is high or inside with a flash mounted on the camera. Both these situations give a very flat and even light which may be ideal for some colour photography but for black and white photography it all too often produces grey, dull and uninteresting photographs.

Learning to control light and use it creatively is an essential skill for a good photographer. When studying a photograph that has been well lit you need to make three important observations concerning the use of light:

1. What type or quality of light is being used?
2. Where is it coming from?
3. What effect does this light have upon the subject and background?

Quality of light

Light coming from a compact point such as a light bulb or the sun is said to have a very 'hard quality'. The shadows created by this type of light are dark and have well-defined edges.

Light coming from a large source such as sunlight that has been diffused by clouds or a light that has been reflected off a large bright surface is said to have a very 'soft quality'. The shadows are less dark (detail can be seen in them) and the edges are not clearly defined.

The smaller the light source, the harder the light appears.
The larger the light source, the softer the lighter appears.

Claire - Lorraine Watson 1992

Claire - Lorraine Watson 1992

A soft light positioned high and to the right gives a flattering and glamourous effect. Note the soft shadows from the glasses.

A single hard light positioned directly above casts deep shadows around the eyes giving a dramatic effect which complements the pose. Note how the hands add to the expression.

Activity 1

Look through assorted photographic books and find some examples of subjects lit by hard light and examples of subjects lit by soft light. Describe the effect the light has on the subjects' texture, form and detail and the overall mood of the picture.

Direction of light

The direction of light decides where the shadows will fall and its source can be described by its relative position to the subject. The light may be high, low, to one side, in front of or behind the subject.

The subject may be lit by a single light source or more than one. This additional light may be reflected back onto the subject from a nearby surface or may be shining directly onto the subject from a second light or be used to illuminate the background.

Broken light - Kevin Ward 1994

Activity 2

Find an example of a photograph where the subject has been lit by a single light source and an example where more than one light has been used. Describe in each the quality and position of the brightest or main light and the effect this has on the subject. In the second example describe the quality and effect the additional light has.

Exposure compensation

In taking a light meter reading of a particular subject you are taking an average value between the light and the dark areas. This meter reading is only accurate where the subject and background are evenly illuminated or have similar tones. It is very important that you 'compensate' or adjust the exposure when your subject has a very bright or very dark background.

If your subject is surrounded by very bright or very dark areas this will cause the light meter to take an inaccurate light meter reading of the main subject. The camera has no way of knowing which part of the viewfinder is the subject you have chosen to photograph, e.g. light behind the subject makes exposure difficult. The camera's meter will be influenced by the light source and indicate an exposure setting that will reduce the light reaching the film. In these situations you have to override the meter unless you want to take pictures of silhouettes.

The Rhondda - Mark Galer 1980

Most cameras take information for the light reading mainly from the centre circle in the viewfinder. When you need to set the exposure for a particular subject either:

1. move in closer so that the face fills the whole frame, and set the exposure, then move back to your chosen camera position and take the shot at the same setting, or
2. point the centre circle of the viewfinder away from the dark or bright area, set the exposure and reposition the camera. Some automatic cameras have a memory lock to help you do this.

Activity 3

Find two photographs where the photographer has shot into the light or included the light source. Explain how the photographer would have gone about taking a light meter reading for these photographs.

Depth of field

You can increase or decrease the amount of light reaching the film by moving one of two controls. By changing the shutter speed (the amount of time the shutter stays open for) or by changing the f-stop (the size of the aperture through the lens).

If you change the aperture, the final appearance of the photograph can differ greatly. This will be the area of sharp focus in the scene, from the nearest point that is sharp to the farthest. This is called **'depth of field'**.

The widest apertures (f.2, f.4) give the least depth of field.
The smallest apertures (f.11, f.16) give the most depth of field.

Try this simple experiment to help you understand depth of field. Place your hand at arm's length in front of your eyes and look at your hand. You will notice the room behind your hand is out of focus. Similarly if you look at the far wall of the room, your hand will appear out of focus. This is called 'shallow depth' of field and is how your camera will see it at the widest aperture. If you progressively reduce the aperture value, known as stopping down, the amount that is in focus gradually increases until you reach f.16 or f.22 when you achieve 'maximum depth of field'. At this aperture it is possible to have both your hand and the wall in focus at the same time. This may not appear obvious at first, because as you look through your camera at f.16 no change appears to have taken place. This is because the lens stays at its widest aperture to allow you to focus even when changing the f-stops. The instant you depress the shutter the aperture closes to the f-stop you have selected. Some cameras have a 'depth of field preview button' located near the lens which allows you to see the true extent of sharpness.

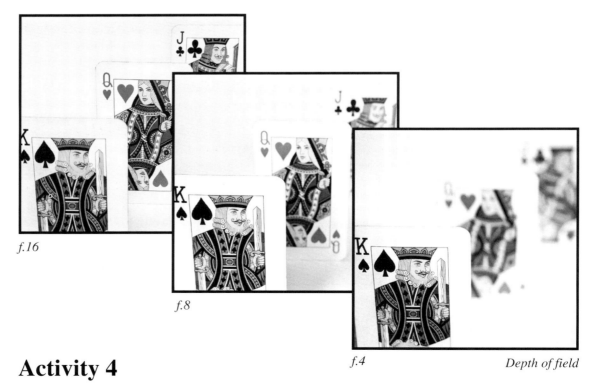

f.16

f.8

f.4   Depth of field

Activity 4

Find four examples of photographs which make use of maximum depth of field and examples which have very shallow depth of field. Describe how the photographer's selective use of aperture affects the subject in each of the photographs you have chosen.

Basic studio lighting

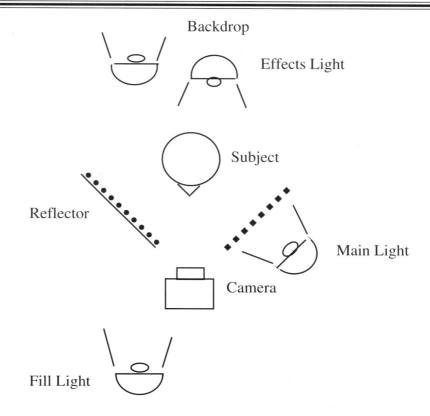

The subject - this may have to be placed at least a metre from the backdrop if shadows are to be avoided.

Main light source - the position is optional depending upon the desired effect required. If deep shadows are created by the main light they can be softened in one of three ways:

1. Reflector - this is used to bounce light from the main light back into the shadow areas.
2. Fill - this is usually a weaker light or one moved further away and is usually positioned by the camera.
3. Diffusion - the hard directional light can be softened at the source either by bouncing it off a white surface or by using tracing paper placed in front of the light itself.

Effects light - this can be shone onto the background to create 'tonal interchange'. This is a technique where the photographer places dark areas of the subject against light areas of the background and vice-versa.

Alternatively the effects light can be shone directly onto the back of the subject so as to 'rim light' otherwise dark areas. Care must be taken to avoid shining the light directly into the camera lens. This is usually achieved by placing the effects light low or obscuring the light directly behind the subject itself.

Practical assignment

Produce a set of six prints that express your feelings towards one of the following titles:

1. Shadows and silhouettes.
2. Broken light.
3. The twilight hours.
4. Face and figure.

Your final work should develop the insight gained from all areas of your research and should include investigations into both natural and artificial light. Try experimenting with window light, studio lights, projected images, light bulbs, candles, torches etc.
If you are having difficulty thinking of a suitable subject try working with another student using each other as 'models'. You may decide to use some other person for the final piece of finished work.

Presentation of work

Research and all contact prints should be carefully mounted on large sheets of paper or thin card no larger than A1. Explanatory notes and comments should be made directly onto this sheet or on paper and glued into place. You should edit your contact sheet including any alterations to the original framing with a chinagraph pencil or indelible marker pen. You should clearly state what you were trying to do with each picture and comment on its success. Contact prints and photographs should be easily referenced to relevant comments using either numbers or letters as a means of identification. You should clearly state how your theme has developed and what you have learnt from your background work and how this has contributed towards your final set of prints.

Resources

1. The Art of Black and White Photography - Pages 15-32 ....John Garrett.

2. Basic Photography - Pages 111-131 ....Michael Langford.

3. Photographic Composition - Pages 70-73 .....Tom Grill and Mark Scanlon.

4. The Complete Kodak Book of Photography - Pages 95-113 ....Hamlyn.

5. Light and Film .... Time Life Books.

Hands 1 - Sanjeev Lal

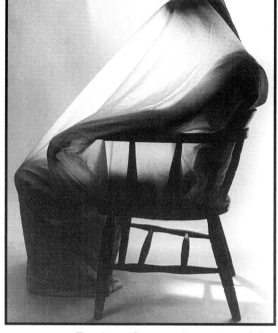

Furniture Constructions - Daniel Cox

Using a single hard light source the student has been careful to remember to frame the subject in a dynamic way. Note how the hands add to the expression of the portrait.

A single light pointing back towards the camera has been placed behind the figure sitting in the chair. The student has rearranged the order of the chair frame, its cover and the sitter.

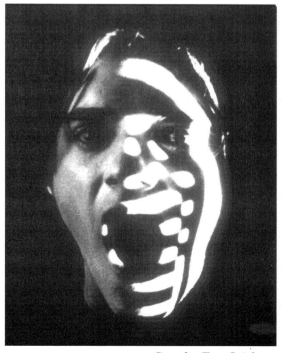

Gareth - Tom Scicluna

Embrace - Rew Mitchell

The contours of the face break the pattern of light being shone from a slide projector positioned to the right of the face.

Hard light, positioned above and left, is used to create this dramatic study of the human figure. The arms are used to create a frame within a frame.

Notes

TIME & MOTION

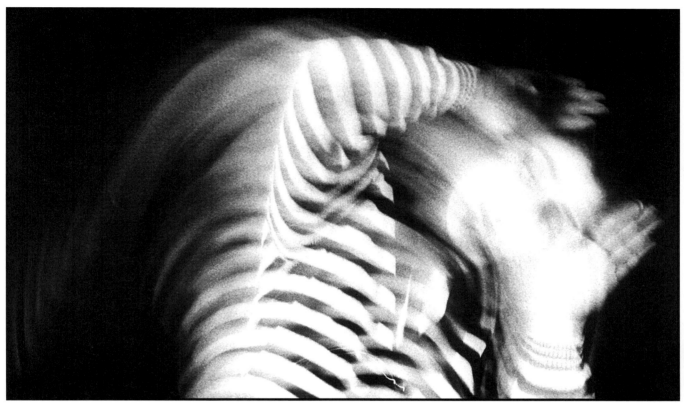

Movement - Renata Mikulik

Aims

These are your aims:
~ To develop an understanding of how a photograph can describe a subject over a period of time as selected by the photographer.
~ To develop an appreciation of how the selected period of time or 'shutter speed' can effect the visual outcome of the print.

Objectives

These are the things you have to do:
~ **Research** - Produce a study sheet that looks at the effect that varying the shutter speed has on a moving subject.
~ **Discussion** - Exchange ideas and opinions with other students on the work you are studying.
~ **Practical Work** - Produce a series of photographic prints that explore a moving subject by using a variety of techniques and shutter speeds.

Notes

Introduction

All photographs are time exposures of shorter or longer duration, and each describes an individually distinct parcel of time.

The photographer, by choosing the length of exposure, is capable of exploring rapidly moving subjects in a variety of ways.

By choosing long exposures moving objects will record as blurs. This effect has been used by many photographers to convey the impression of motion. Although describing the sense of movement it could be said that they do not describe a great deal about the object itself in motion. Possibly the most interesting thing photography can do with movement is to destroy it.

Gare St. Lazare - Henri Cartier-Bresson

The decisive moment

Henri Cartier-Bresson in 1954 described the visual climax to a scene which the photographer captures as being the 'decisive moment'.

In the flux of movement a photographer can sometimes intuitively feel when the changing forms and patterns achieve balance, clarity and order and when the image becomes for an instant a picture.

Activity 1

Look at a Henri Cartier-Bresson photograph and discuss why you think that shooting at the decisive moment has added to the picture's quality.

High speed photography

By freezing thin slices of time, we can explore a new beauty. A fast shutter speed may freeze a moving subject yet leave others still blurred. The is dependent on the speed of the subject matter.

Activity 2

Find an example of a photograph where the photographer has used a very fast shutter speed and describe the subject matter including the background. Discuss any technical difficulties the photographer may have encountered and how he or she may have overcome them.

Discuss what has happened to the depth of field and why.

Discuss whether the image gives you the feeling of movement stating the reasons for your conclusion.

Wimbledon - Nigel Bramley 1994

Slow speed photography

Rodeo - Ernst Haas

The photograph opposite shows the effect that can be achieved by using a slow shutter speed to capture a fast moving subject. When the shutter speed slows down still further ther is the risk that the subject matter is no longer recognisable.

Activity 3

Find a photograph where the photographer has used a slow shutter speed and describe the subject matter including the background. Discuss any technical difficulties the photographer may have encountered and how these may have been overcome.

Discuss what has happened to the depth of field and why.

Discuss whether the image gives you the feeling of movement stating the reasons for your conclusions.

Photographic techniques

Panning

This technique describes the action of following the movement with the camera. You must begin to track the subject before you press the shutter release and follow through or continue to pan once you have made the exposure.

Very long exposures

For this technique you will need to mount the camera on a tripod and fire the shutter using a cable release. You should select the smallest aperture possible on your camera lens, e.g. f.16 or f.22, and use a slow film such as 50 ISO. The shutter speed can be further extended by the addition of a light reducing filter such as a polarising filter.

Rew - Chris Gannon

Slow shutter speeds

For this technique you can either hold the camera or mount the camera on a tripod. By using shutter speeds slower than those recommended for use with the lens you are using you can create blur from moving objects. If you are using a standard lens try shooting a moving subject at speeds of 1/30th, 1/15th, 1/8th and 1/4 of a second. If the camera is in a fixed position on a tripod the background will be sharp but if the camera is panned with the subject the background will produce most of the blur. Try shaking the camera whilst exposing to increase the effect even more.

Fast shutter speeds

For this technique you need to select a wide aperture. Some telephoto and zoom lenses only open up to f.4 so you will need to attach a lens that will allow you to open up to f.2.8 or wider. Once you have selected the aperture take a light meter reading of your subject. If your indicated shutter speed is only up to 1/250th then you must do one or both of the following. Either increase the amount of illumination reaching the subject and/or use a faster film.

Multiple exposures

This can be done in a number of ways:

1. After loading the film but before the exposure, rewind the film using the rewind crank so that there is no slack in the film (do not depress the rewind button at this stage). Place some masking tape or similar over the rewind crank to prevent the film moving. If the background is illuminated or you are shooting outdoors you will need to reduce the exposure, i.e. reduce the aperture by 1 stop for 2 exposures and 2 stops for 4 exposures. After each exposure and before you wind on to cock the shutter for the next exposure you must depress the film rewind button which is usually on the camera base. If you are shooting against a black background in a studio and the subject is moving away from its original location it will not be necessary to reduce the exposure.

Movement - Alexandra Rycraft

2. Multiple exposures can also be achieved using flash equipment. This is best achieved in a dark studio using a black background. The camera can be mounted on a tripod and the shutter fired using a cable release. The camera's shutter speed dial should be set to B and the cable release kept depressed until the desired number of flashes have been fired. Allow the flash to recharge after each firing. The flashgun will need to be fired manually and you should also ensure you have set the correct aperture on the lens as indicated by your flashgun or flash meter. See the section about working with flash in this study guide for more information.

Zooming

For this technique you need to use a lens that can alter its focal length, i.e. a zoom lens. The camera should be mounted on a tripod and a slow shutter speed selected, e.g. 1/15th, 1/8th or 1/4 of a second. The effect of movement is achieved by making the exposure whilst altering the focal length or zooming the lens either in or out. The subject does not need to be moving for the effect to work.

Manipulation in the darkroom

A stationary subject or one that has been frozen using a fast shutter speed can be blurred by moving the print slowly during the final seconds of exposure in the darkroom. Movement or time passing can also be achieved by using David Hockney's 'Joiner' technique.

Flash photography

Flash is an extremely bright light of exceptionally short duration. It is important to note that there are only three main kinds of flash although there are many different makes and sizes. They are manual, automatic and dedicated and they are discussed in more detail below.

Manual flash

A manual flash will give out a fixed amount of light each time it is fired. You calculate the correct exposure by first estimating the distance between the flash and the subject. Looking this distance up on the back of the flashgun will give you the correct f-stop which you have to set on your camera lens. The disadvantage of this type of flash is that it can be slow and imprecise.

Automatic flash

An automatic flash is usually more expensive but has the advantage of having a thyristor which reads the amount of light reflected from the subject and cuts short or 'quenches' the flash when it has calculated the film has been correctly exposed. In this way the subject distance can vary without the photographer having to change the aperture set on the camera lens.

Dedicated flash

Dedicated flash guns are designed to work with specific cameras. The camera and flash communicate more information through additional electrical contacts in the mounting bracket of the gun. The TTl metering system of the camera is used to make the exposure reading instead of the thyristor. In this way the exposure is more precise and allows the photographer the flexibility of using filters without having to alter the settings of the flash.

Working with a flash gun

1. Check that the film speed has been set on the flash.
2. Check that the flash is zoomed to the same focal length as the lens. This may involve pulling the head of the flash forward or back.
3. Check that the shutter speed on the camera is set to the correct speed (usually 1/60th of a second or slower).
4. Check that the aperture on the camera lens is set to that indicated on the flashgun. On dedicated units you may be required to set it to an automatic position or the smallest aperture.
5. Check that the subject is within range of the flash. The flash will only illuminate the subject correctly if the subject is within the two given distances indicated on the flashgun. If the flash is set incorrectly the subject may be overexposed if too close and underexposed if too far away.

Slow sync flash

Slow sync is a technique where, by mixing daylight at a slow shutter speed and a high speed flash which freezes movement, you create a picture which is both blurred and sharp. This is a popular technique to record action shots that have a feeling of movement and yet still give a recognisable image of the subject.

Bowling Alley - Mark Galer

For this technique follow the procedure below:

1. Load the camera with 100 ISO film.
2. Set the camera shutter speed to a slow setting, e.g. 1/15th, 1/8th or 1/4 second.
3. Take a light meter reading and set the correct f-stop on the camera lens. If you are unable to select an f-stop small enough to obtain the correct exposure (you are letting a lot of light reach the film by using such a slow shutter speed), try reducing the amount of light by using a polarising filter, neutral density filter or slower speed film. Make sure you compensate for these actions on the flash.
4. Set the aperture that is on your lens on the flashgun's control panel. If you are unable to select the same aperture that is on your lens adjust the shutter speed and lens aperture until you can. If you wish to darken the background so that the subject stands out more simply increase the shutter speed over that recommended by the camera's light meter.

NB. If you have few working apertures on your flashgun it may be quicker to select the smallest one then select the same aperture on the camera lens and finally set the shutter speed. If you find that the shutter speed after following the above is still too quick to create blur try shooting when the ambient light is low, e.g. dusk.

Practical assignment

Produce a series of six prints that explore a moving subject by using a variety of techniques and shutter speeds. Your final presentation sheet should demonstrate that you have developed a theme from your initial investigations and that you have chosen appropriate techniques for your subject matter.

A possible title for your set of prints could be:

1. Sport
2. Dance
3. Vertigo
4. The fourth dimension

Your work should:

a) Make use of different shutter speeds, including some work at or near the slowest and fastest speeds possible with the equipment you are using.
b) Demonstrate an understanding of what is meant by the Decisive Moment.
c) Show that you have considered both lighting and composition in your work.

Presentation of work

Research and all contact prints should be carefully mounted on large sheets of thin card no larger than A1. Explanatory notes and comments should be made directly onto this sheet or on paper and glued into place. You should edit your contact sheet including any alterations to the original framing with a chinagraph pencil or indelible marker pen. You should clearly state what you were trying to do with each picture and comment on it's success. Contact prints and photographs should be easily referenced to relevant comments using either numbers or letters as a means of identification. You should clearly state how your theme has developed and what you have learnt from your background work and how this has contributed towards the final set of prints.

Resources

1. <u>Photographic Composition</u> - Pages 80 - 85 .... Tom Grill & Mark Scanlon.

2. <u>Henri Cartier-Bresson</u> .... Aperture Masters of Photography.

3. <u>The Camera</u> - Pages 88 - 96 .... Time-Life on Photography.

4. <u>Ernst Haas</u> .... Bryn Campbell.

5. <u>Action Photography</u> .... Bryn Campbell.

6. <u>Motion & Document-Sequence & time</u> ... James Sheldon & Jock Reynolds.

Any books including the work of Eadweard Muybridge, Gjon Mili and Harold Edgerton

Leapfrog - Melanie Sykes

Sparks - Clair Blenkinsop

The camera was mounted on a tripod and set to f.11 for a 1/8th of a second exposure.

The camera's shutter was held open on the B setting in a dark studio whilst a student traced the outline of the model with a sparkler.

kick - Mark Galer

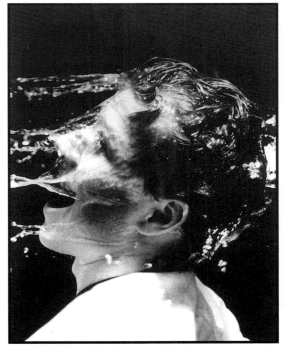

Splash - Claire Ryder

This slow sync flash effect was created by setting the camera's shutter to 1/15th second, using a 500watt tungsten light and the flash placed slightly to one side of the subject.

Two 1,000 watt halogen lights were used to obtain this fast 1/500th second exposure. Back lighting was essential to illuminate the water against the dark background.

Notes

Advanced Study Guides

PHOTOMONTAGE

Agoraphobia - Chris Gannon

Aims

Your aims are:
~ To develop an awareness of how visual elements interact with each other and with text to convey specific messages.
~ To develop an awareness of the techniques employed by individuals in the media in order to manipulate photographic material.

Objectives

These are the things that you have to do:
~ **Analyse** a range of assembled photographic images from both commercial and non-commercial sources.
~ **Produce** artwork to demonstrate how assembled photomontages can communicate a specific message.
~ **Evaluate** the effectiveness of your own work.

Notes

Introduction

When we look at a photograph we read the images as if they were words to see what information they contain. If the photographic image is accompanied by words they can influence our decision as to what the photograph is about or even alter the meaning entirely. The photographer can manipulate the message to some extent by using purely photographic techniques such as framing, cropping, differential focusing etc. This controls which information we can or cannot see. Some artists and photographers have chosen to increase the extent to which they can manipulate the information contained in a photograph by resorting to the use of paint, chemicals, knives, scissors and electronic means to alter what we see. The final result ceases to be a photograph and becomes a 'photomontage'.

Definition

A photomontage is an image that has been assembled from different photographs or from a single photograph that has been altered. By adding or removing information in the form of words or images the final meaning is altered. The resulting photomontage may be artistic, commercial, religious or political.

Destruction - Darren Ware

History

Photomontage is almost as old as photography itself. During the 1850s, several photographers and artists attempted to use photography to emulate the idealised scenes and classical composition that was popular in the Pre-Raphaelite painting of the period. Photographs during this time were restricted by the size of the glass plate used as the negative in the camera. Photographic enlargers were not common, so the images had to be contact printed onto the photographic paper. By using many negatives to create a picture the photographer could increase the size of the final image. This also released the photographer from the need to photograph complex sets using many models. The technique of 'combination printing' allowed the photographer to photograph each model individually and then print them on a single piece of paper masking all the areas of the negative that were not needed.

Historical examples

A famous combination print of the Victorian era is called 'Two Ways of Life' by Oscar Gustave Rejlander, produced in 1857. As many as 30 negatives were used to construct this image. The picture shows the moral choice between good and evil, honest work and sin.

Another photographer who worked with this technique was the artist Henry Peach Robinson. Robinson first sketched his ideas for the final composition on a piece of paper and then fitted in pieces of the photographic puzzle using different negatives. One of his most famous images is the piece titled 'Fading Away' which was constructed in 1859 from five negatives. The image depicts a young woman dying whilst surrounded by her family.

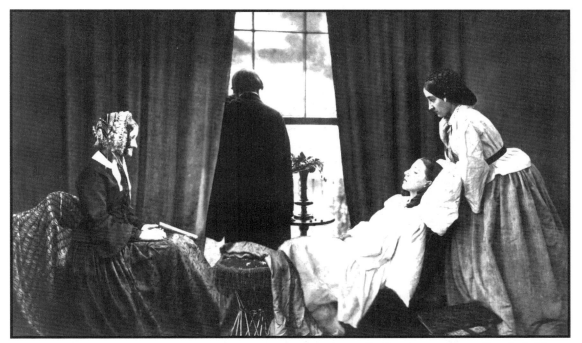

Fading Away - Henry Peach-Robinson

Historical developments

Both Robinson and Rejlander were heavily criticized by the fine art world for their work in this field but the technique was popular with the public and did survive. Commercial photographic studios in the 1860s started making their own composite photographs. They were made by cutting up many photographs, usually portraits of the famous or of beautiful women, and gluing them onto a board. This was then re-photographed and the resulting prints sold as souvenirs or given away as promotional material.

Activity 1

Briefly discuss why you think painters of the Victorian period might have criticised these early photomontages and why the pictures were popular with the general public.

This technique of piecing together separate images to create one picture is again very popular with both artists and the media. What two reasons can you think of for this revival of an old technique.

Political photomontage

The technique of photomontage was not widely used again until the Cubists and the Dadaists in the 20th century experimented with introducing sections of photographs and other printed material into their paintings. A Dadaist by the name of John Heartfield further developed this use of photography and is now commonly attributed as the founder of political photomontage.

John Heartfield was influenced by Dadaism and the newly emerging socialism in the Soviet Union. Sergei Tretyakov, a Russian born writer, showed Heartfield how a collection of facts carefully edited could convey a significant message. John Heartfield was already deeply involved with politics and he now saw how the graphic use of assembled photographs, combined with the mass production printing techniques of the time, could reach a wide audience with his political messages.

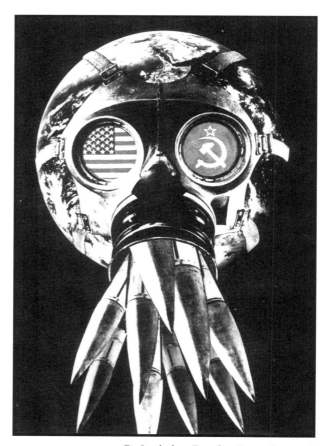

Defended to Death - Peter Kennard

The techniques that John Heartfield developed in the 1920s and 1930s are still used by contemporary artists such as Klaus Staeck and Peter Kennard. In the 1980s Peter Kennard used photomontage to draw attention to the escalating arms race, his most famous works being associated with the organisation CND (the Campaign for Nuclear Disarmament). Both artists have found photomontage an ideal medium to communicate social and political injustice to the public in an immediate and effective way. Through photomontage the public are asked to question the media images they see everyday, thus raising their visual literacy.

Techniques of political photomontage

1. **Text:** Text included with most photomontages sets out to reinforce the message of the photographer or artist. The words can remove any possible ambiguity that the viewer may find, draw the attention to the conflict in the images or contradict the images entirely thus establishing a satirical approach to the work.

2. **Recognising the familiar:** The viewer is meant to recognise familiar images, paintings, photographs, advertising campaigns etc. that the photomontage is based upon. The viewer is drawn to the differences from the original and the new meanings supplied by the changed information. This technique is often used in contemporary photomontage works. The technique is to make the familiar unfamiliar, e.g. 'Cruise Missiles' Peter Kennard, 'Mona Lisa' Klaus Staeck.

3. **Contradiction:** In many photomontages we see images that contradict each other. Introduced parts of the photomontage may be inconsistent with what we would normally expect to see happening in the image thus questioning the original point of view. The contradiction can also be between what we see is happening and the text. Our curiosity to establish a coherent meaning in both cases is raised.

4. **Contrast:** To gain our attention individual components of the photomontage may be in sharp contrast to each other, e.g. wealth and extreme poverty, tranquillity and violence, happiness and sadness, industrialisation and the countryside, filth and cleanliness etc.

5. **Seeing through the lies:** Many photomontages invite the viewer to look behind the surface or see through something to gain greater insight into the truth. Windows that open out to a different view, X-ray photographs that reveal a contradiction are all popular techniques.

6. **Exaggeration of scale:** By altering the scale of separate components of the photomontage the artist can exaggerate a point, e.g. a photomontage entitled 'Big business' may include giant, cigar smoking, industrialists crushing smaller individuals under their feet.

7. **Figures of speech:** A very popular technique of the photomontage artist is to visualise figures of speech, e.g. puppet on a string, playing with fire, house of cards etc. The artist may play upon the viewer's acceptance of these as truth or use them as a contradiction in terms, e.g. 'The camera never lies'.

Activity 2

Find two examples of political photomontages that are either from a historical or contemporary source. Discuss in what context they have been produced and how effective you think they communicate their intended message.

Discuss the techniques that have been used to assemble the examples you have chosen and offer alternative ways that the artist could have put over the same message.

Photomontage in the media

Advertisers and the press have embraced the new technology, that makes image manipulation easy, with open arms. It is fast, versatile and undetectable. Advertising images are nearly always manipulated in subtle ways to enhance colours, remove blemishes or make minor alterations. We are now frequently seeing blatant and extensive manipulation to create eye-catching special effects. We do not get unduly agitated or concerned over these images because advertising images have always occupied a world that is not entirely 'real'.

To manipulate or not to manipulate

We start to move into more questionable territory when images we take to be from real life, perhaps of real people, are altered without our knowledge. The top fashion models we see on the covers of magazines like Vogue and Vanity Fair for instance, we believe to be real people, especially when they have been taken out of the advertising context. We expect models to have near perfect features and complexions, but it would come as a shock to many sections of the public to realise that these images are frequently altered as they are not quite perfect enough for the picture editors of the magazines. The editors do not distinguish between the images that appear on the cover of the magazine and the images that are sent to them in the form of advertisements. Minor blemishes to the model's near perfect skin are removed, the colour of their lipstick or even their eyes can be changed to suit the lettering or the colour of their clothes. Many women aspire to these models and their looks and yet these people don't exist. It is important to remember that a publication is a product, whether it is a fashion magazine or a daily newspaper, and as such editors may be more interested in sales than in truth.

Most editors are at present using their own moral codes as to when, where and to what extent they will allow manipulation of the photographic image. Most editors see no harm in stretching pictures slightly so they can include text over the image or removing unsightly inclusions. What each editor will and will not allow can vary enormously. Occasionally the desire for an image to complete or complement a story is very strong and the editor is tempted to overstep their self-imposed limits. A fabricated image can change the meaning of a story entirely or greatly alter the information it contains. It is possible to construct a visual communication where none may otherwise have existed.

Activity 3

As editors exercise their ever increasing power over information control, what limits would you impose on them as to the extent to which they can manipulate the photographic image?
Devise a series of guidelines that will control the release of images that have been constructed for media use so that the public are aware as to the extent of the manipulation.

Activity 4

You have been commissioned to produce a series of posters that aims to increase awareness about the effects of image manipulation. The posters are intended as part of a study pack for A level media studies students. The pack will include a booklet of photographs and supporting text about this important subject.

The posters should illustrate the following statement:

A photograph, whether it appears in an advertisement, a newspaper or in a family album is often regarded as an accurate and truthful record of real life. Sayings such as 'seeing is believing' and 'the camera never lies' reinforce these beliefs. The information we see in photographs however, is often carefully selected, so what we believe we're seeing is usually what somebody would like us to see.

By changing the information we can change the message.

The publishing company of the study pack envisages a series of photomontages or image manipulations that will cover the implications of image manipulation in all areas of the printed media.

Produce **ONE** finished poster in colour or monochrome. You may use a mixture of black and white with coloured text. With the finished poster you should submit a set of layouts and your preliminary sketches. The work that you submit should include the following:

1. The final poster should be supported with at least four different design layouts at a reduced scale. These layouts should demonstrate a variety of creative approaches to this assignment and include the text that would accompany the image. You should have considered designs for posters dealing with both press and advertising manipulation.

2. The final choice of poster should be supported with a 200 word 'rationale' that explains how the poster will raise general awareness to the issues of image manipulation. The rationale should also include the techniques you have used to communicate this message.

3. Text included in the final design could be one of the sayings 'seeing is believing' or 'the camera never lies' which should conflict with the image that you produce or it could be some copy of your own that reinforces the message conveyed by your image.

4. The size of the final poster should either be A2 or A3 and include the chosen text in a size that can be easily read from 3 metres.

Photomontage in art

Around the time that John Heartfield was developing the language of political photomontage, another artist, Laszlo Moholy-Nagy, was using photography in conjunction with drawn geometric shapes. The effect was to cause a visual conflict between the print viewed as a factual record of three dimensional reality and the print as a two dimensional surface pattern. This important work ensured that many contemporary artists would continue to use the camera as a creative tool. Some have found 'straight photography' limited as a medium for visualising their own personal ideas or abstract concepts within a single image. Image manipulation and photomontage are two ways that photography can be used by the artist to communicate in more complex ways.

Surrealism

Surrealism followed Dadaism and photography was found to be an ideal tool and so became directly involved in this movement as it had been in Dadaism.

Surrealist images are constructed from the imagination rather than from reality. Many of the works by Surrealist painters such as Salvador Dali and Rene Magritte appear as fantastic dream sequences where familiar objects are placed in unusual settings. Parts of the image may be distorted in scale or shape and the impossible is often visualised. Inspired by the work of these surrealists Angus McBean in the 1930s, produced a whole series of photomontages of actors and actresses. The photographs are a strange mixture of fantasy and portraiture where the character is often surrounded by props from their current play.

Audrey Hepburn - Angus McBean

Joiners

David Hockney has used photomontage in many of his works. Individual photographs have been placed together to create a larger photographic collage which Hockney has termed a 'Joiner'. The number of prints in each Joiner may vary depending on how much information Hockney feels he needs to explore in the subject matter. In some of his Joiners he has photographed more than one side of an object or person and then placed the individual images together to create a single picture. This technique was used by cubist painters such as Georges Braque and Picasso who show many facets of objects on a single canvas. Hockney has taken this one stage further and presented the subject at different moments in time. The photographic instant has been lengthened in some of Hockney's Joiners to become a study of a 'happening', different views from different times during the same event. A picture of a card game is no longer a record of that game at one particular moment but rather a study of the whole game.

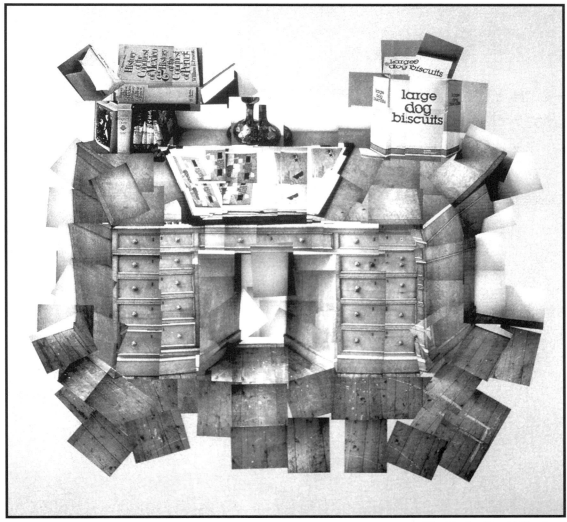

The Desk - David Hockney

Activity 5

Find two examples of photomontages that are either from a commercial source or from a fine art background. Discuss in what context they have been produced and what techniques they share with political photomontages.

What messages, if any, are communicated through these photomontages and how effective do you think they are? Consider different ways that each image could have been tackled by the artist and come up with an idea for 'another in the series.'

Practical assignment

Produce one 16" x 20" photomontage or a series of smaller montages in response to one of the following titles:

a) **Phobia** - an abnormal or morbid fear of something.
b) **Metamorphosis** - a change of form, character or conditions.
c) **Decay** - decompose
d) **Dream world**
e) **Strange but true**
f) **Fast food**
g) **Fashion**

Additional information

Your work should include:

1. A detailed description of how you have photographed each individual element of the montage and the sources for 'found' images that you may have had to use (images taken from the printed media).
2. A description of how the final montage has been assembled.
3. A research sheet including your own contact sheets and ideas.

Resources

1. <u>Photomontage - A Political Weapon</u> .... David Evans & Sylvia Gohl
   Gordon Fraser London 1986

2. <u>Photomontage Today</u> - 35 minute video .... Arts Council of Great Britain

3. <u>In Our Own Image</u> - The Coming Revolution in Photography .... Fred Ritchin
   Aperture Foundation New York 1990

4. Books including work by the following artists and photographers:

   - <u>Historical Photomontage</u> - Oscar Gustave Rejlander & Henry Peach Robinson

   - <u>Political Photomontage</u> - John Heartfield, Klaus Staeck & Peter Kennard

   - <u>Surrealism</u> - Rene Magritte, Salvador Dali, Laszlo Moholy-Nagy & Angus McBean

   - <u>Contemporary Artists</u> - David Hockney

Eating Cars - Zara Cronin

Space & Pavement - Darren Ware

Notes

Notes

DISTORTION

Distortion - Ashley Dagg-Heston

Aims

Your aims are:
~ To extend personal creativity through improved technique.
~ To explore the limitations of photographic materials and equipment through experimentation and exploration.

Objectives

These are the things you have to do:
~ **Analyse** various techniques of photographic distortion as used by historical and contemporary photographers.
~ **Produce** artwork that explores a number of techniques and processes and produce a series of prints that communicate a personal theme or idea.
~ **Evaluate** the effectiveness of your work.

Notes

Introduction

All photographs distort reality to a lesser or greater degree. The very act of converting three-dimensional reality into a two-dimensional print is a distortion in itself. The simplest of photographic techniques used by the photographer and editor can manipulate and distort the information that the photograph puts forward to the viewer. This can be achieved in the following ways:

- the act of framing composes the visual elements within the frame and decides which of the visual elements we can or cannot see in the final photograph.

- the viewer can be led through the print in a systematic way through the use of line, tone and colour.

- the choice of lighting can effect the mood of the image, to reveal or disguise form and texture.

- use of shallow depth of field called 'diffferential focusing' can guide the viewer to individual elements within the frame.

- movement can be explored or the illusion of movement created through the controlled use of shutter speed and flash lighting.

- visual elements can be removed or added after the photograph has been taken to change the meaning of the image.

- the use of text can be used to clarify or contradict meaning.

The limits of distortion
Photographic techniques control the way we read information from photographs in many ways, yet most people accept the accuracy of the photographic medium in recording the subject matter in front of the lens. This study guide is designed to give you the opportunity to discover the limits of distorting this information using both the equipment and the materials commonly used in the photographic process.

Activities
a) Complete **one** technique from each of the five numbered categories in this study guide. Use subject matter you feel is appropriate to the visual effect. Record the procedures you have used accurately so that the process can be perfected at a second or third attempt.
b) Research both art and media sources for examples of photographic distortion and make notes as to how you think this effect may have been achieved.

NB. The final practical assignment requires that you work within a theme (as with previous assignments). It is advisable that you consider this theme whilst working through the activities section of this study guide. This will ensure that your research work is seen as appropriate.

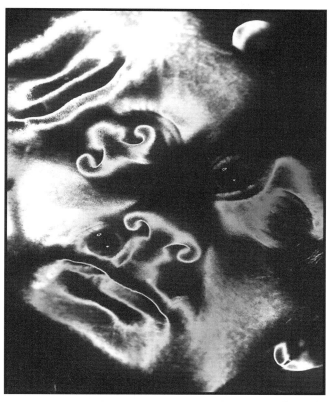

Faces - Paul Heath

This print is a double exposure which has then been solarised whilst processing the print.

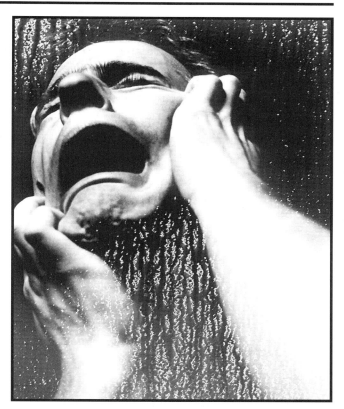

Scream - Lynsey Berry

This image was distorted by spraying the developer from a plant sprayer whilst the developing tray was tilted.

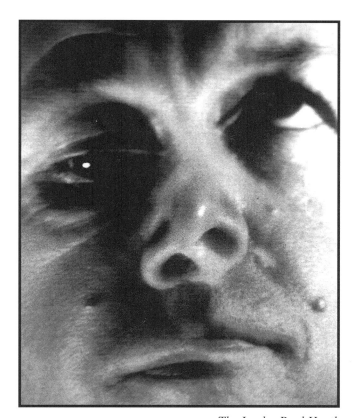

The Look - Paul Heath

This image was created in the camera by a double exposure. Only one side of the face was illuminated for each exposure.

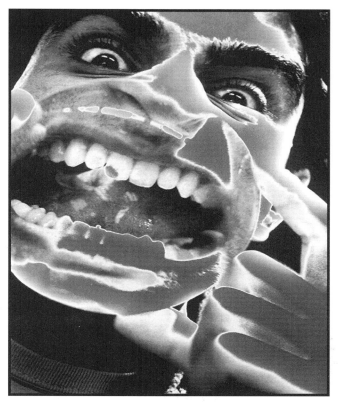

The Mouth - Gareth Neal

This image is a combination of distorting the face using a magnifying lens and then solarising the final print.

Distortion using the camera

Lens distortion

Telephoto or long lens - A telephoto lens compresses, condenses and flattens three dimensional space. Subject matter appears to be closer together. This effect is most noticeable when the lens has exceeded the focal length of 135mm.

Very long lenses are expensive but extreme effects can be explored by attaching a teleconverter to a shorter focal length lens. A focal length of 400mm or greater may be achieved in this way without paying large sums of money. The problems encountered by using this technique are maximum apertures that may be no better than f.8. A fast film will be necessary in these situations.

Wide-angle lens - A wide-angle lens exaggerates distances are scale is distorted. Subjects close to the lens looks larger in proportion to their surroundings. Subject matter in the distance looks much further away. The overall effect is one of 'steep perspective'.

A wide-angle lens with a focal length of 28mm or shorter is recommended to explore this technique. The closer you move to the subject the greater the distortion.

Zooming - For this technique you need to use a lens that can alter its focal length, i.e. a zoom lens. The camera should be mounted on a tripod and a slow shutter speed selected e.g. 1/15th 1/8th or 1/4 second. The effect is achieved by altering the focal length during the exposure. The subject does not need to move for the effect to work.

Depth of field - Very short depth of field is achieved by using one or more of the following techniques: a) long focal length lens, b) wide aperture, c) moving closer.

Wide depth of field is achieved by using a combination of one or more of the following techniques: a) short focal length lens, b) small aperture, c) moving further from subject.

Refraction and reflection

Refraction - This is the action of light being bent or deflected as it passes through different media such as glass and water. Look at the swimmer photographed by André Kertész. Try photographing through patterned or textured glass, special filters, clear filters with Vaseline smeared on them, water etc. in front of the lens.

Reflection - When light is reflected off smooth surfaces which are curved we get a view of the image distorted. Look at the nudes produced by André Kertész.

The mixing of reflections with the view through plane glass can also produce interesting effects, as can mirror images introduced into the picture. Possible sources for such images include chrome items, reflective foil, shiny black cars etc.

Lighting and film

Artificial light - Try using unusual lighting conditions such as car headlights or portable video lights to illuminate your subject. Experiment mixing tungsten and flash or by placing opposite coloured filters over the camera lens and a flashgun. Try experimenting with coloured filters using black and white film (a red filter turns a blue sky virtually black).

Infrared film - Infrared is a part of the light spectrum that is invisible to the human eye but can be recorded on special film. Skies appear dark and green vegetation appears to glow. This film needs to be used in conjunction with a deep red filter and must only be handled in complete darkness.

Lith film - This is a very high contrast film which can be loaded directly into the camera and exposed at 6 ISO. The film can be developed in special lith developer so that hardly any grey tones remain.

Point blank - Philip Budd

Depth of field

The photograph above draws the viewers attention to the barrel of the gun. The student has used a mixture of telephoto lens, close proximity to the subject and a wide lens aperture to create this photograph. This is a technique known as differential focusing and is an example of pushing the equipment to its limits.

Distortion in the darkroom

The negative

Little can be done to gain any worthwhile effects during the processing of black and white negatives. The following can be attempted but the effects are unpredictable.

a) **Reticulation** - this is a process designed to damage the grain structure of the negative by subjecting the film to extreme temperature changes during processing. Replacing the stop bath with very hot water and then ice cold water before fixing the film can achieve reticulation. Most modern films are very resistant to this type of abuse or treatment.

b) **Solarisation** - this is a process of fogging the film to white light during the development stage of the negative. The resulting negative, if the process is carried out correctly, will appear half positive and half negative. The effect is more commonly achieved in the printing stages due to the unpredictable nature of the process and is often called pseudo-solarisation or sabatier. If you are going to attempt it using film, try using a high contrast film such as 'lith' film in the camera and concentrate on subjects with bold patterns and that have been lit with a hard directional light source.

The flash of light required to fog the film during the development stage must be of the correct intensity. Try experimenting with short pieces of film loaded onto a film spiral. Remove the developing tank lid in the darkroom (with the red safelight switched off) shortly before half the developing time has elapsed. Place the tank on the baseboard of an enlarger underneath the enlarging lens. Expose the film whilst it is still submerged in the developer to 1 second of light with the enlarger lens on a small aperture. The fogging must occur when the film is only half way through the development stage. Once the film is fogged no further agitation should be carried out until the development stage has been completed. If the resulting negative is too dark after it has been fixed reduce the aperture further or move the head of the enlarger higher on the column. If no noticeable effect can be seen increase the exposure.

c) **Extreme grain** - to achieve very large grain try using hot developer for a reduced amount of time and/or the use of Kodak recording film.

d) **Cross processing** - this is a technique whereby a colour transparency film is processed in chemicals designed to process colour negative material (C41 process). The result is an interesting shift in colours.

e) **Distortion of the film emulsion** - it is possible to mark or manipulate the film emulsion in a variety of ways. Any manipulation of the negative surface is usually permanent so it is recommended that several identical shots of the same subject are taken so that individual techniques can be perfected without the loss of a good shot. The film emulsion can be scratched using a sharp instrument such as a compass or marked with pens and paint. Special opaque paint called 'photo opaque' can be used on film surfaces. This will mask any areas that you do not wish to print. Negatives are constructed from a celluloid layer which can be distorted with chemicals or heat. black and white emulsion layers are very resistant when dry but colour transparency emulsion layers will bubble when heat is applied.

Tone elimination

Tone elimination reduces the number of tones or shades of grey in an image. The image can be reduced to just two tones - black and white-or to three or more. This technique is used in screen printing where only one tone, or colour, can be printed at a time.

i) Selecting an image - Choose an image which has a strong impact and a good tonal range.

ii) Preparing the materials - You will need lith film (either sheet film or 35mm) and lith developer. A contact printing frame will also be necessary if you are using 35mm film (ensure that the glass is clean). The film can be handled in darkrooms that use red safelights.

iii) Making a test strip - Enlarge the original negative onto the sheet film or contact print the strip of negatives onto lith film, emulsion to emulsion (the emulsion side of lith film is lighter in tone). Make a series of timed exposures as if you were making a test strip for a print.

iv) Processing - Process the lith film in a tray with lith developer at 20°C for 2-3 minutes (lith developer deteriorates rapidly once mixed). Wash and fix until clear.

v) Making a positive - Choose the best exposure and repeat the process. Wash and dry the positive well. If you require more than one tone you will need to make positives of varying exposures. The third image opposite was made using three different exposures.

vi) Making a new negative - Repeat the previous process using the positives instead of the original. Wash and dry the negatives well.

vii) Printing the tone elimination - Retouch any dust marks on the lith negatives using photo opaque, a fine paint brush and a light box prior to printing. If you are using only one lith negative to make your print follow your usual printing procedure. If you are printing from several lith negatives you will need to register each negative by using registration marks. Mark important parts of the image with a pen on a piece of paper inserted into the printing easel. Register each negative to these marks before re-inserting the photographic paper. Each exposure will be of the same length, the combined exposures creating the darkest tone in the shadow areas.

Normal print

A two tone print

A four tone print

The print

a) Negative prints - These can be produced from any dry print and are often produced from solarised prints (see the work of Man Ray). First make sure that the light from the enlarger lens covers the baseboard and set the enlarger lens to the brightest aperture. Place a fresh sheet of unexposed printing paper, emulsion side up, on the base board. Place the original print face down onto this unexposed printing paper. Place a clean sheet of glass onto the two sheets of paper to ensure close contact or use a purpose built contact printing frame. Make a test strip to find out the correct exposure. Exposures will be longer than if you had used a negative because the light has to pass through the top print to expose the paper underneath. Process the test strip and the resulting negative print as normal.

b) Solarisation - This is a technique which gives the print the appearance of being both negative and positive at the same time. It is achieved by subjecting the print to a brief second exposure during the development stage. The lightest tones of the print are affected most by this second exposure thus altering the normal tonal values of the print. A white line or halo appears between the first and second exposures. See page opposite.

c) Selective developing and fixing - The processing chemicals can be applied to some of the areas of the printing paper leaving others to fog or remain undeveloped. Many imaginative ways can be found to apply the chemicals or prevent their contact with the print surface. These include spraying, painting or dripping the chemicals and by coating areas of the print with hand cream, Vaseline or masking fluid which can be removed before fixing.

d) Movement of printing paper - The printing paper can be twisted, bent or dragged on the enlarger baseboard for a varying length of the exposure to distort the image or give the illusion of movement.

e) Toning - Many chemicals are available to tone photographs that have already been processed. These include sepia (a warm brown tone), blue and copper toner. The prints should be soaked for half an hour in water. Blue toning a print will darken it, so it may be advisable to start with a print a little lighter than usual.

f) Colouring - Paints, inks and dyes can be applied directly to the finished print. Gloss and resin coated papers are more difficult to work with than matt and fibre based papers and will generally only accept acrylic paints.

g) Sandwiching negatives - This is achieved by placing two negatives together in the enlarger carrier (emulsion to emulsion). Experiment with using one negative of a silhouette and one negative of a surface texture or pattern. If exposures are too long try using negatives that are slightly underexposed.

h) Multiple exposures - This can be achieved by exposing different negatives onto the same piece of printing paper or turning the paper around and re-exposing the paper using the same negative. If a subject has been photographed against a bright background this will produce dark areas on the negative. When the negative is printed areas of the print will be unexposed, allowing a second exposure to be made into this area. This technique will require some planning to be effective.

Making a solarised print

i) Selecting an image - This should be a bold image with a strong pattern. The image could contain blocks of differing tone or strong graphic lines.

ii) Make a test strip - Using grade 4 or 5 makes a high contrast test strip. If the negative will not give you a high contrast print choose another negative.

iii) Make a print - Choose the best exposure from the test strip and reduce the exposure by 25-30%. Make 2 or 3 prints but do not process them. Store them in a box or black plastic bag.

iv) Preparing for the second exposure - Remove the negative from the enlarger. Make sure that the light from the enlarger lens covers the baseboard of the enlarger either by raising the enlarger head or altering the focus. Cover the baseboard with paper towels and place a developing tray half full of water onto these. The light from the enlarger must cover all of the tray. Switch off the enlarger.

v) Make a second test strip - The aim of making this second test strip (without the negative) is to find a middle grey tone. This tone will indicate the time required for the second exposure. Set the enlarger timer.

vi) Half process the print - Process one print that you made earlier but remove it from the developer when it has received only half the developing time. The image should only be half the correct density. Slide the print gently into the tray of water and let the surface settle.

vii) Second exposure - Switch on the enlarger to give the print its second exposure.

viii) Final processing - Return the print to the developer and process normally. Fix and wash as usual.

ix) View the print - View in daylight conditions and alter the time of the second exposure if needed.

Normal print

Solarised print

Negative of solarised print

Distortion on the computer

To produce images using a computer you need to:

- have access to a flatbed scanner or a 35mm negative/transparency scanner
- be able to save or store your images onto a floppy disk
- have access to a software package that allows electronic manipulation
- have access to an inkjet or laser printer

1. **Scanning** - Negatives or positives can be scanned (the information converted into digital information the computer can understand). The scanning software will usually allow some manipulation at this stage but it is usually restricted to brightness, contrast and colour balance. Scanning an image will generally produce a file which contains a great deal of information. If you do not want to wait for a very long time for images to print out or need to store several images onto a floppy disk make sure the image scans at a fairly low resolution. Unless you have access to some very sophisticated and expensive equipment images produced from a computer will not be able to match the quality of an image produced using photographic paper. Black and white images containing no more than 300K of memory and printed at 150dpi will however produce acceptable images. It is common to find that an 8 x 10 photograph will produce a file size which will not fit onto a floppy disk. If this happens and you do not wish to loose the quality by reducing the file size it is possible to compress the information when saving the image.

2. **Filters** - Computer software packages designed for electronic imaging purposes usually come with a set of filters that allow special effects to be applied to an image. These include effects which resemble many that can be created in the camera and darkroom and some that are unique to computers.

3. **Retouching** - The imaging software will allow each pixel or group of pixels to be changed to ones of the operator's choosing. The software allows pixels to be chosen in a variety of ways from a palette (a selection of tones and colours) or from other areas of the image. These chosen pixels can then be applied to areas of the image using a variety of tools. These tools are designed to apply the chosen tones or colours in localised or broad areas depending upon the effect required. The retouching technique allows the operator to remove unwanted details or blemishes, improve contrast or colour saturation in localised areas or change colours or tones entirely.

4. **Montage** - The imaging software allows the operator to assemble an image from components from different images. Sections of an image can be cut out and pasted down onto a different image. The imported image can be moved to a precise location on the new image before permanent insertion. The size, colour or tone of the imported image can be changed to match the new image so that the completed manipulation is undetectable. This technique allows photographs that would be very difficult, expensive or impossible to produce with conventional equipment feasible.

Image manipulation

The images opposite demonstrate how copying and pasting can be used to manipulate the original image. The first image shows a man with a broken nose and one half of his face in shadow. It was decided to duplicate the left hand side of the photograph to produce the manipulated image at the bottom of the page.

Cut, copy and paste

The electronic imaging software has several tools which allow the operator to cut out, or copy sections of an image and paste them in a different place or in a different image entirely.

The tools above are:

Top left - Rectangular marquee tool
Top right - Elliptical marquee tool
Bottom left - Lasso
Bottom right - Magic wand

A selection using the marquee tool is made by first clicking on the tool you require and then by clicking on the image and dragging the mouse until the area of image that you require is indicated by a dotted line.

The lasso is used to draw around a section of the image whilst the magic wand is placed on a specific area of the image and clicked. The magic wand will select all the pixels that are either the same or are similar to the ones where the magic wand was placed.

The selection is then either copied (leaving the original image intact) or cut, using the command from the Edit menu at the top of the screen.

The section of image selected can be modified in many ways before or after it has been moved to the new location. In this example the section being moved is finally flipped so that we then view a mirror image of itself. When the new section is correctly positioned it is finally pasted down (using the command from the Edit menu). Any areas that don't quite match can then be retouched to produce a manipulation that is very difficult to detect.

The original image

The selection is moved

The selection is reversed & deselected

Practical assignment

Choose **one** assignment only. All work is to be mounted on A1 or A2 card and should be supported with at least two sheets of research and background work. The background work should consist of the work carried out in the 'activities' section of this study guide plus evidence of how your ideas have developed for your chosen theme.

1. A publishing company has commissioned you to produce a set of photographs that will advertise their forthcoming book titled <u>Beyond Reason</u> one of which will be used on the dust jacket of the book.
The book is about psychic phenomena and it is envisaged that the photographs will use special effects to set the mood for the articles in the book.

2. You have been commissioned by a successful photographic studio to establish and promote a specialised area of photography. The area is to be known as <u>The Dirty Tricks Department</u>. This will be a mixture of specialised photographic effects together with original creative thinking.
 The promotion is to be a leaflet containing a number of images and you may wish to include the title as part of the presentation.

3. Choose one of the following themes from which to work. Think very carefully about the many ways you can approach your chosen theme. Do not choose the most obvious definition of the word unless you have considered alternatives.

Sensation
Force
Transitions
Rituals
Faith
Masks
Dreams

4. Produce a set of promotional photographs that would be suitable for the pressure group Amnesty International. The images you produce should promote a particular aspect of their work that you feel strongly about.

5. Express in photographs one or more of the following emotions. You may consider using props and/or text to convey the emotion effectively.

Anger
Fear
Love
Hate
Joy

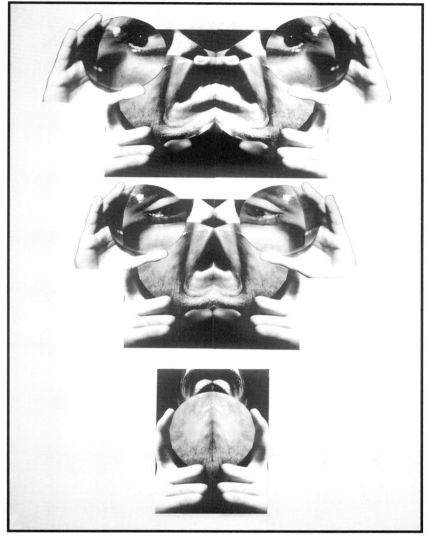

Distortion 2 - Gareth Neal

The student has created a sequence of distorted images. A magnifying lens has been used to distort sections of the human face whilst taking the shots. These have been further distorted by using montage techniques to piece them together with mirror images of themselves. The sequence shows the original image at the top disappearing into itself.

Resources

The Print, Time Life Publications .... images created in the darkroom
Man Ray ... solarisation, negative prints, multiple exposures and photograms
Bill Brandt - Nudes 1945-1980 .... use of wide angle lens distortion
André Kertész - Nudes .... distorted reflections
Ernst Haas .... use of slow shutter speeds
Classroom Photography - Ilford (London) .... creative use of materials
Creative Darkroom Techniques .... Kodak
The Workbook of Darkroom Techniques .....John Hedgecoe/Mitchell Beazley

Notes

SELF IMAGE

Positive and Negative - Faye Gilding

Aims

Your aims are:
~ to develop an awareness of the links between self image and personal identity.
~ to develop an awareness of symbolism and visual codes of practice.

Objectives

These are the things that you have to do:
~ **Analyse** a range of categories that define or divide people and the visual symbols connected with these categories.
~ **Produce** artwork that explores one or more aspects of your personal character.
~ **Evaluate** the effectiveness of your work.

Notes

Introduction

We are judged as individuals by our appearance, our actions and our status within society. These judgements, which influence the way we are treated, contribute to our awareness of 'self' and our self image.

The images we choose to represent ourselves are important ways of establishing our identity. In photographs we might find evidence of the way we conform, share similarities and fit in with society, e.g. uniforms, fashion, dress code etc. In the same photograph or in a different photograph we might find evidence of what makes us an individual, a unique person within society, e.g. expression, make up, jewellery, important personal objects, locations and other people.

The images we put forward of ourselves may vary depending on how we would like to be perceived by others. An individual may like to be seen as smart, serious and powerful in a business setting, but casual, friendly and relaxed in a family setting. The images we choose to frame or keep in our albums however will tend to portray positive aspects of our self image, e.g. happiness, success and security. The images we choose could be seen to bear a resemblance to the images we see each day of people used by the media industry to promote their products.

Classifications

As individuals we tend to seek out groups that we feel comfortable in. These groups contain individuals that are similar, in many respects, to ourselves and together the group forms a collective identity. From this collective identity, whether it be a team, family group, club or fraternity, we draw a part of our self image that is important to our own identity. Members of large collective groups adopt visual forms of recognition, e.g. items of clothing, body decoration or adornment (scarves, rings, tattoos, haircuts etc.). Along with the visual forms of recognition, the individual is expected to adopt the customs, rituals or ceremonies that serve to unite the group in a common mutual activity, e.g. worship, singing or chanting, dancing etc.

The adolescent years and years of early adulthood can be very turbulent and explosive for some individuals as they strive either to 'fit in' or 'break out' of some of the categories they were brought up in, or labelled with. As individuals establish new identities conflict is caused leading to insecurity and anger on both sides.

The struggle to recognise and define identity is one of life's most difficult tasks.

Activity 1

What follows is a list of categories that serve to define and divide us. Make a personal list of the categories you belong to, or have conflict with and visual symbols that are associated with each. Aspirations and expectations should also be listed.

1. Age/generation - This can dictate levels of independence afforded to the individual.
2. Gender - Our sex affects the way we are treated by some individuals.
3. Race - Prejudice and perceived racial superiority can affect hopes and aspirations.
4. Religion - Moral codes and ideologies that serve to guide, unite and divide.
5. Class or caste - A classification that an individual may never escape from.
6. Personality - Extrovert and introvert are classifications of personality.
7. Intelligence - We are often segregated by educational establishments using IQ.
8. Political persuasion - Left wing or right wing. Each carries its own ideology.

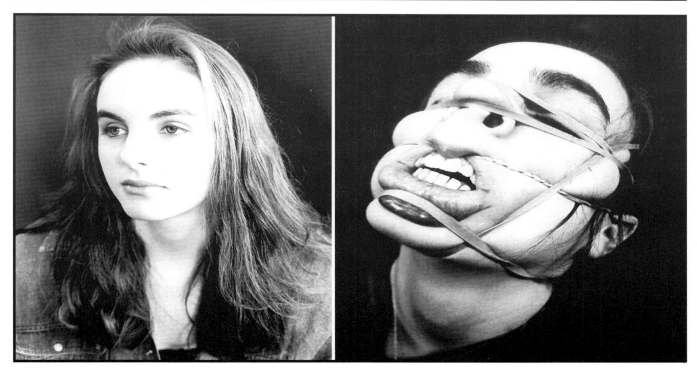

Beauty and the Beast - Catherine Burgess

Images of beauty

The photographs above are of the same student. Each image represents a different aspect of the same character. The image on the left represents how the student would most like to be seen, positive, calm and physically attractive. The image on the right represents how the student would least like to be seen, disfigured, in pain and physically repulsive. Many people edit images of themselves before they are passed around or placed into a photographic album. Some people will actually destroy images of themselves that they do not feel portray them in the best possible light. We are drawn to the image on the right because of its visual impact but we are shocked to learn that the image on the left is of the same student. We do not expect that a single individual has the capacity to be both beautiful and ugly, yet in reality we all feel these reactions to our own image at different times, depending on our state of mind, our physical well being and the manner in which the camera has captured our likeness.

Activity 2

Find images in the media which have been used to represent attractive and unattractive aspects of the human face. Examine and record carefully the photographic techniques used to accentuate both these qualities, drawing up a list that relate to the images you have found.

List the physical characteristics that we have come to admire in both the male and female face and write 100 words in response to the following questions:

1. Do you believe that media images or public opinion is responsible for the characteristics of beauty becoming universal stereotypes?
2. Do you believe any harm can be done by people admiring media images of glamourous models?

Fashion Statement - Thomas Scicluna

Images of conformity

The photomontage above shows the same figure dressed in many strange ways. Each figure is accompanied by a bar code as seen attached to the products we buy in the supermarket. The orderly lines of figures contrast with the main figure in the foreground who is wearing his underpants outside his trousers and who stares straight at the viewer in an unnerving manner.

The student is making the statement that fashion manipulates the young. Fashion rather than being an individual attempt at personal expression is a commercial way of pressurising people to conform (no matter how silly the conformity looks). The main figure looks like he has just recognised this fact and is not amused.

Boxes - Housewife - Julia McBride

The photomontage above expresses the students awareness of social pressures to conform to the image and behaviour of a typical housewife. The student is seen to be breaking out of the 'box' in which she feels she may be placed at some future date.

Activity 3

Consider some of the social pressures that you think may shape your behaviour and personal image. List the images most commonly associated with the categories or 'boxes' you have already listed in Activity 1. How have you responded to social pressures to conform by adopting an appearance that relates to the categories that you feel you have been placed in or have chosen.

Practical assignment

Using photography and other media put together a composite image or series of images that communicate how you see yourself in harmony or conflict within the social structure of which you are a part. The assignment may put forward one aspect or several concerning your self image and it is envisaged that you will explore not just the positive ones.
Your design sheets or sketch book should show evidence of the development of a variety of ideas or approaches to this assignment and together with the finished piece of work they will form an integral part of the assessment material.
Possible starting points for practical work concerning self image:

1. Reflections:
    a window into the soul?

2. Different aspects or sides to our character could be illustrated through conflicting, contrasting or changing images:

- Negative and positive, strong and weak, child and adult etc.

- Different aspects of our character viewed simultaneously or consecutively.

- Triptychs, installations, constructions, e.g. cubes, cylinders, mobiles etc.

- Our dreams and aspirations versus the reality we inhabit.

- A recognition of the sources of individual pressures within society to con
    form or adjust our behaviour and appearance.

- Stereotypes - racial, gender etc.

Resources

Media images representing age, class, gender and race.

The family photograph album.

Personal collections of photographs.

The photographic work of Jo Spence.

The photographic work of Cindy Sherman.

Facades - Daniel Shallcross

Daniel Shallcross produced this image by breaking a mirror into many pieces on the floor and photographing the reflection of a single individual. Due to the shallow depth of field the actual fragments of the mirror cannot be seen. Daniel was working on the idea of an individual possessing more than one identity and how the image may be a mask to the true identity.

Reflection - Alison Brown

Alison has used a highly reflective piece of card which has been bent to produce this dramatic effect. The model and their reflection can be viewed at the same time but the reflection is a distorted view of reality. Alison has produced an image which explores how our self image can be distorted by our levels of confidence and by the recording methods used.

Notes

READING THE IMAGE

Mark Galer

Aims

Your aims are:
~ To develop an understanding of why and how photographic images are constructed.
~ To develop an awareness of how the photographic image can be manipulated to communicate specific messages.

Objectives

These are the things you have to do:
~ **Analyse** a range of photographic images from the printed media, documentary and advertisements.
~ **Produce** artwork individually and in teams to demonstrate how photography can be used to communicate different messages.
~ **Evaluate** how effective the media is in its aims.

Notes

Introduction

A photograph, whether it appears in an advertisement, a newspaper or in a family album is often regarded as an accurate and truthful record of real life. Sayings such as 'seeing is believing' and 'the camera never lies' reinforce these beliefs. In this study guide you will learn that the information we see in photographs is often carefully selected by the photographer or by the editor so that what we believe we're seeing is usually what somebody else would like us to see. By changing or selecting the information we can change the message.

The concepts of time, motion and form that exist in the real world are accurately translated by the photographic medium into timeless and motionless two-dimensional prints. Photographers, editors and the general public frequently use photography to manipulate or interpret reality in the following ways:

1. Advertisers set up completely imaginary situations to fabricate a dream world to which they would like us to aspire.
2. News editors choose some aspects of an event, excluding others, to put across a point of view.
3. Family members often choose to record only certain events for display in the family album, excluding others, so that we portray the image of the family as a happy and united one.

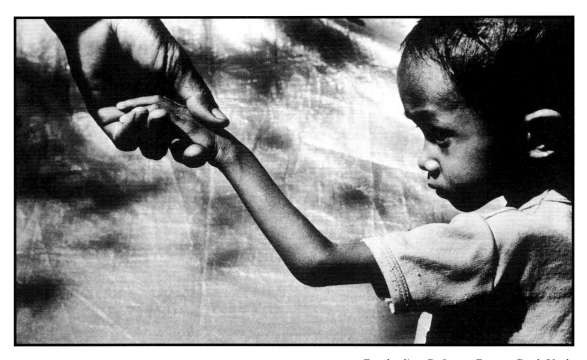

Cambodian Refugee Camp - Burk Uzzle

Photography is a powerful media tool capable of persuasion and propaganda. It appears to offer truth when in reality it can portray any manipulative or suggestive statement. The camera may record accurately but it is people who choose what and how it records. A photograph need only be sufficiently plausible so that it appears to offer the truth.

Manipulation techniques

Framing and cropping - This technique is used to include or exclude details that may change the meaning of the photograph. The photographer, by moving the frame, or the editor by cropping it, defines the content of what we see. By framing two facts it can create a relationship where none may otherwise exist. By excluding a fact it may break one, e.g. an old woman sitting on a park bench feeding the birds might appear lonely if we have moved the frame slightly to exclude her granddaughter playing nearby.

Editing - A photographer chooses a subject matter that he believes to be important and selects the decisive moment to take the picture. In order to present a personal point of view the editor may then decide to select only one aspect of the photographers work.

Captions - A caption may give the picture a moral, social, political, emotional, or historical meaning that emphasises the intended message.

Focusing - The amount we see 'in focus' can vary and draw our attention to a specific part of the photograph.

Vantage point or angle - The photograph may be taken from above, below or at eye level. Photographs taken from a low vantage point tend to increase the power and authority of the subject whilst those taken from above tend to reduce it.

Lighting - This can change the atmosphere or mood of the scene, e.g. soft and subtle, dramatic or eerie.

Film type - Black and white or colour. The contrast and colours may vary.

Subject distance - The closer we appear to be to the subject the more involved we become and the less we discover about the subject's environment or location.

Lens distortion - By using a telephoto lens detail and depth can be compressed, condensed and flattened. Subject matter appears to be closer together. By using a wide-angle lens distances are exaggerated and scale is distorted. Things that are close to the lens look larger in proportion to their surroundings than they really are. Things in the distance look much further away. An estate agent may use a wide angle lens to make a room look bigger.

Composition - This is a technique which is used to describe how lines, shapes and areas of tone or colour are placed within the picture frame to attract or guide our attention, e.g. diagonal lines, whether real or implied, make the picture more dramatic and give us a sense of movement. Due to the strong colour contrast, a person wearing a red jacket in a green field will instantly draw our attention.

Photographic categories

Documentary - A factual record.
Advertising - The promotion of a product.
Art - A display of skill and/or creative expression.

Activity 1 - coverage

Study a range of magazines, tabloid and broadsheet newspapers. What percentage of each paper's surface area do photographs cover? Break down your findings, separating the images into the three categories mentioned earlier and present the information diagrammatically.

Activity 2 - analysis

A) Remove a selection of photographs complete with captions from the media sources you have used in activity 1. After studying each photograph prepare a table, as in the example below, using the following headings:

- Category
- Subject matter
- Techniques used
- Implied messages

In describing the subject matter you will need to consider:

- Who is the main figure in the photograph
- How they look
- Who else appears and how they react to the main figure
- What else appears in the photograph e.g. objects, location, setting.

Category	Subject matter	Photographic Techniques	Implied message
Advert for beauty product	Woman well-dressed in the grounds of a large country house. Handsome well-dressed young man looking on.	Warm colours, softly lit low vantage point, focus on woman. Woman placed centrally in foreground.	Affluence, glamour, sexual admiration, happiness, contentment.

B) 20 years ago John Berger wrote in *Ways of Seeing* that advertising showed us images of ourselves, made glamorous by the products it was trying to sell. He claims the images make us envious of ourselves as we might become, as a result of purchasing the product. Berger also believes that advertising makes us dissatisfied with our present position in society.

Look at the images you have collected. Which of the following are we likely to envy or admire in each individual image:

- Power
- Prestige and status
- Happiness
- Glamour
- Sexuality

Do any of your advertisements appeal to different emotions other than admiration and envy? With advertising becoming increasingly sophisticated and less blatant do John Berger's claims still ring true?

Activity 3 - captions

In this activity you will look at how captions reinforce the message implied in a photograph. Captions that accompany photographs can be:
- Explicit - stating something very clearly as fact
- Implicit - suggesting that something is true

For this activity you need to choose photographs and their captions from both the advertising and documentary categories. Rewrite the captions that accompany each to give a different point of view or bias. You may like to think of captions that change the moral, political, emotional or historical meaning of the photograph. The caption you choose may even change the category in which the photograph first appeared.
Present your work and discuss how you have changed the meaning of the photographs and how effective the new messages have become.

Activity 4 - photomontage

Photomontage is a technique where separate photographs are combined to create new meanings. The interaction between the new elements creates a new meaning.
Below we see the unintentional and incongruous juxtaposition of two posters in the street. "An entirely new meaning is made in the distance that exists between the glamour and eroticism of western fashion and the economics of survival in the third world." (Peter Kennard - Photomontage Today).

Mike Wells 1980

Using some of the photographs you have collected together cut out various elements from each and rearrange them to make new images with new messages. The sort of new images that work best are where strong contrasts are placed together, e.g. unemployment queues standing next to advertisements for luxury merchandise.

101

Activity 5 - Group practical

For this practical activity you should split up into groups of three or four. Your group will have a broader perspective if the members are not all male or female.
The things you will need for this activity are as follows:

- 35mm SLR camera
- 24 exposure roll of film
- location
- assortment of props

If you intend to work in a studio setting you may consider the following:

- studio lights
- tripod
- background or backdrop

Using appropriate photographic techniques as discussed earlier, take one photograph for each of the following categories:

- Advertising
- Documentary

Repeat the exercise using one or more inappropriate techniques to create a satirical photograph for each of the categories. Allow six frames for each individual in the group.

Important

Discuss in detail each shot before you commit yourself to taking the picture. You might like to consider a possible caption for each shot before planning your photographs. Prepare for any technical difficulties that you think you may encounter, e.g. working with a tripod if you are shooting indoors without a flash.
You have probably already noticed how many photographers fill the frame with their subject matter in order to make a bold statement. Avoid standing too far back in your own shots unless you have a specific reason to do so.

Activity 6 - editing

This activity requires that you work in pairs to act as a newspaper editor and photojournalist. Your objective is to produce a short news story that is either:

- Biased
- Unbiased

Use titles, photographs and captions. Text is optional but you may consider pasting down some text to create a realistic mock-up of the finished article.
The editor may choose not to tell the photographer which stance he or she is going to take on the subject. The photographer may choose to manipulate the range of images he supplies to the editor.
Your teacher will give you advice on the range of topics you may choose from, or you can make your own suggestions.

Activity 7 - The family album

A) In this activity you will be looking at the way the family is portrayed both in the media and in our own family albums.

Collect some media photographs that portray the family and analyse them using the criteria you used in activity 2:

- Category
- Subject matter
- Photographic techniques
- Implied message

Apply the same criteria to the photographs that appear in a typical family photograph album. Pay particular attention to the way the photographs have been edited. What aspects of family life were edited out or simply not recorded in the first place? Do you think the family album is a true representation of family life? How useful or damaging would it be to alter the type of events that are considered worthy of being placed in the family photograph album?

B) Write a short essay evaluating your findings. Discuss how you think media images have affected you, your family and the general public. Try to be as honest as you can.

Resources

The following resources are suggestions only.
You or your teacher may wish to add to this list.

Images from the press
1. Newspapers (both 'tabloid' and 'broadsheet').
2. Sunday supplement magazines.

Recommended reading
3. S Sontag. On Photography. Penguin Books.
4. J Berger. Ways of Seeing. Penguin Books.

Images with accompanying text
5. C Steele-Perkins. About 70 Photographs. Arts Council of Great Britain.
6. M Langford. Basic Photography. Focal Press.
7. J Szarkowski. Looking at Photographs. Museum of Modern Art, New York
8. B Dutton. Media Studies. Longman.
9. The Photographer's Gallery - London. Learning Resources Projects.
10. P Kennard. Images for the End of the Century. Journeyman Press.

Video
11. Photomontage Today. Peter Kennard.

Personal photographs
12. Family photograph albums.

Notes

Teaching Resources

TEACHING RESOURCES

Mark Galer

Aims

The aim of this guide is:
~ To offer resource material for teachers.

Objectives

The resources enclosed in this guide:
~ Offer a structured approach to student learning.
~ Offer information and advice about planning to install a darkroom.
~ Offer record sheets for the monitoring of student progress.
~ Offer a sample test paper for monitoring annual student achievement.

Notes

Darkroom design

Basic requirements

Constructing a darkroom does not have to be expensive or complicated. An available room can be modified quickly and effectively if the following are present:

1. A sink with running water
2. An electricity supply
3. An extractor fan to supply adequate ventilation

Preparing the room

The dry surface - A dry surface is needed where the enlargers have access to a power supply. In an ideal environment each enlarger has its own power supply just above the work surface. Each enlarger needs about one metre of available work surface and these can be partitioned off up to the height of the enlarger being used. Although this is useful to prevent the light from one student's enlarger fogging the paper of another, it is by no means essential if students take care to switch off their enlarger whilst removing and replacing the negative carrier.

The wet surface - A wet surface for the processing of printing paper is also required. If the money is available a flat bottomed PVC sink long enough to contain the four processing trays on a rack placed inside the sink is ideal but not essential. The first three trays can be placed directly onto a washable surface and the final wash tray can be placed in the adjacent sink where it can have access to running water via a flexible hose. The water is allowed to overflow this final tray in the washing process. Precautions can be taken to avoid flooding caused by small pieces of photographic paper or paper towels blocking the plug hole by inserting a small plastic tube into the plug hole thereby raising the water outlet to the sink.

The entrance - The most convenient entrance to a darkroom that is used by several people is a light trap painted matt black. This allows people to enter and leave the darkroom without the need for others to put away light sensitive materials. The alternatives to a light trap are double doors or a revolving door which although less convenient will take up less space.

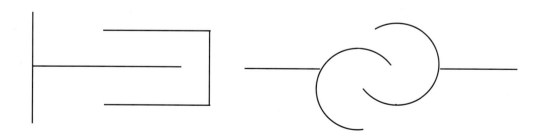

Two examples of a darkroom light trap

Ventilation - A light trap will allow free circulation of air but doors must be provided with a light tight grill. The extractor fan must also be light proof. Any materials used to prevent direct light entering the darkroom should be painted with matt black paint.

Safelight - The most convenient form of safelight available are special red fluorescent tubes that fit directly into the existing fittings. If more than one fitting is available on separate switches one can be left with the original tube to provide bright illumination for preparation of chemicals and clearing up at either end of the day. Safelights can also be bought which screw into existing bayonet type fittings. When choosing the type of safelight you will also need to consider whether you will be handling certain types of film which can also be handled in safelight conditions. If you are not sure as to which safelight you will need for the long term you are advised to choose red lights rather than orange as these will allow students to use all black and white materials that can be handled in safelight conditions.

Towel dispenser - Students will need the use of towels to dry hands after washing and mop any spillage of water or chemicals.

Essential equipment

Enlargers - Enlargers should be bought that can use filters to alter the contrast of multigrade paper. These can either be equipped with a multigrade head, colour head or have a filter draw. When purchasing the enlarger enquiries should be made as to whether it comes with a 50mm lens suitable for enlarging from a 35mm negative.

Developing trays, printing tongs and rubber gloves - The trays should be slightly larger than the printing paper you intend to use. The printing tongs and rubber gloves are essential to allow students to transfer prints which are wet with chemicals between trays.

Mixing cylinders and funnel - These are necessary for diluting the concentrate and returning diluted chemicals to containers for storage.

Eye protection - Students must wear eye protection if they are expected to pour chemicals.

Clock - A large wall mounted clock with a sweeping second hand should be mounted on a wall that is easily visible by all the students working in the darkroom. This will be of use for timing operations where there is a lack of individual timers and will also make sure that students do not overrun photographic sessions.

Non-essential equipment

Enlarger timers - These are useful and accurate but again are not essential. In-line switches on the power cables to the enlargers can be fitted to switch the enlarger light on and off. Students can be taught to use exposure times longer than 10 seconds and count accurately by speaking the same word between each number, e.g. one-potato-two-potato-three etc.

Printing easels - These are used to hold printing paper flat and in the right position on the baseboard. They are however a luxury and are easily damaged. Modern resin coated printing paper will lie flat without the use of an easel. A separate board with a white sticky back plastic oblong the same size as the printing paper can be used to frame the projected image and provide positioning for the photographic. If borders are required on the photographic paper a window mount can be cut from card and used together with a baseboard which will register both the paper and the window mount.

Supplementary equipment - Focus finders, rotary trimmer, drying rack. These make life easier in the darkroom but can be added after the darkroom is already established.

Controlled test

Centre name_____   **Date** _____

Centre number _____   **Time allowed** _____

Instructions

Your completed work must include contact sheets and research work as well as your finished presentation piece. You should give clear evidence of how you have approached your work and how your ideas have developed during the course of the test. Any experimentation into style and technique should be recorded clearly on the research sheets. The final work should be mounted on A1 or A2 card.

You are encouraged to make full use of the range of techniques and skills covered by the course wherever appropriate.

Answer <u>ONE</u> question only.

1. **Manual labour**   You have been commissioned by the council to present a visual documentary about manual labour. Your piece could focus on a man, woman, group or team. Your work should emphasise the interaction between people and their surroundings, with particular reference to the supporting objects and images associated with their occupations and pastimes.

2. **Theatre workshop**   Produce a set of promotional photographs for a theatre workshop that show the company are encouraged to develop a wide variety of acting styles. Your photographs should demonstrate a variety of moods and emotions using one or more people. It is recommended that you stage the lighting and poses specifically for the camera. The moods and emotions might include happiness, sadness, curiosity, contemplation, boredom, excitement, friendliness, hostility, arrogance, delight, fear, satisfaction, anticipation, anger, peace, concentration, uncertainty, and frustration.

3. **Pressure group**   Produce a set of promotional photographs that would be suitable for use by a pressure group, e.g. Greenpeace, Friends of the Earth, Campaign for Nuclear Disarmament, Anti Nazi League or one of your own choice. The images you produce should promote the principal objectives or symbolism related to the particular group.

4. **Illustration**   Produce a set of images which illustrate a poem, book extract, song lyric or newspaper article. You may incorporate the text into the finished piece of work.

5. **Surreal story**   Produce a sequence of images that communicate a surreal short story. The source of your imagery could be a recent dream or daydream that you may have experienced.

6. **Hands** Photograph hands in expressive postures or engaged in interesting activities. You may photograph one hand by itself, both hands of one person or the hands of several people together. Do not include a full face with the hands, though part of a face is acceptable.

7. **Still life**   Produce a set of still life photographs that explore a range of old objects, things that are worn out from age or use. The final work can be either abstract or commercial.

8. **Shadows** Photograph an object (or part of it) along with its shadow, exploring how the shadow can add visual interest to an image.

Progress report

Student Name .... Year .... Group ....

Date .... Time............. Place...............................

Assignment feedback

Creativity ....

Composition ....

Presentation ....

Exposure ....

Processing ....

Focusing ....

Contrast ....

Attendance .... Punctuality .... Deadlines ....

General progress

Creative approach to design activities ....

Management of time ....

Technical competency ....

Appropriate use of the study guides to aid research ....

Points raised by student.

....

....

Points raised by tutor

....

....

Negotiated statement on progress

....

....

Necessary action agreed to be undertaken

....

....

Signature of student ....

Signature of tutor ....Date ....

Work sheet

Student Name ....Year .... Group ....
Assignment ....

Research

1. What is it you like about the images you have been looking at? ....
....
....
2. What techniques has the photographer used? ....
....
....

Preliminary work

1. Which images do you like from your first contact sheets? ....
....
....
2. How could you improve upon these images? ....
....
....

Plans and ideas

1. What elements from your research can you use in your own work?...............
....
....
2. What ideas do you have to develop a theme? ....
....
....

Organisation of final shoot

Location .... Time and date ....
Estimated time needed for shoot ....
Permission or tickets required .... Contact name ....
Equipment needed ....
Props needed ....

Summer project

Introduction

The word 'Summer' can conjure up many different images: strawberries and cream, hot lazy days, rain soaked day trippers, excited holidaymakers, bored school children, cricket on the village green, discarded ice-creams in the sand, littered beaches etc.

Assignment

In this assignment you must record one aspect of your summer this year, something that has attracted your visual curiosity. Do not necessarily choose a stereotypical view of summer, i.e. a picture postcard shot of the beach or family snap shots.

The images that you choose should relate with each other in some way and portray a theme or aspect of summer that you would like to convey. The photographs may be abstract, documentary shots from real life or shots that have been set up to convey a message.

You may work in black and white or colour and will be expected to have taken at least two rolls of film. It is not expected that you will process or print your own images over the summer. You may have your films processed and printed commercially. If you decide to work in black and white you will find that the film XP2 can be processed quickly and cheaply through most photographic high street minilabs.

Edit your prints down to six images and present them on one sheet of card no larger than A3 in size. The deadline for this project is the first day of term.

Assessment criteria

Your work will be assessed using the following criteria:

1. Arrangement and composition of subject matter within the frame.
2. Sympathetic use of viewpoint and creative use of shutter speed.
3. Visual clarity of idea and theme.
4. Presentation

Notes

Glossary

Analyse To examine in detail.

Ambient/available light The natural or artificial continuous light that exists before the photographer introduces additional lighting.

Aperture A circular opening in the lens that lets light through to the film.

Blurred An image or sections of an image that are not sharp. This can be caused through inaccurate focusing, a wide aperture or a slow shutter speed.

B setting The shutter speed setting B allows the shutter to stay open as long as the release remains pressed.

Bounce Lighting that is reflected off a surface before reaching the subject.

Burning-in Giving more exposure time to selected areas of the print.

Cable release A cable that allows the shutter to be fired without shaking the camera when using slow shutter speeds.

Close-up lens A one element lens that is placed in front of the normal lens that allows the camera to be focused closer to a subject.

Close down A term referring to the action of making the lens aperture smaller.

Composition The arrangement of shape, tone, line and colour within the boundaries of the image area.

Contact print A print which has been made by placing negatives or positives in direct contact with the printing paper.

Contemporary photographer A photographer living or working at the present time. The term contemporary may however be used when referring to someone who lived or worked at the same time as another.

Context of the image The facts that relate to when and where the image was taken.

Contrast The difference in brightness between the darkest and lightest areas of the image. A very high contrast image is one which would reveal deep blacks, bright whites and few if any tones of grey. A very low contrast image is one which would have many tones of grey but few, if any, areas of black or white.

Cropping The act of trimming or masking off unwanted edges of the image to compose the subject matter.

Decisive moment The moment when the arrangement of the moving subject matter in the viewfinder of the camera is composed to the photographer's satisfaction and the shutter release is pressed.

Dedicated flash A flash gun that uses the camera's own light meter to judge the subject has received the correct exposure.

Density In photographic film the density refers to how dark the image is and how much light it allows through.

Depth of field The distance between the nearest and furthest points of the subject matter that are acceptably sharp at a particular aperture setting.

Diagonal A slanting line that is neither horizontal nor vertical.

Differential focusing A technique to ensure that only a specific part of the subject matter is in focus so that the viewer's attention can be guided to this point.

Diffused light Light from a large light source that creates shadows that are not clearly defined. A cloudy sky in which the sun can't be seen will give a diffused light.

Dissect To cut into pieces. The edge of the frame can dissect a familiar subject into an unfamiliar section.

Dodging Reducing the exposure to selected areas of the print to alter the tone without effecting the rest of the print.

Edit Select images from a larger collection to form a sequence or theme.

Effects light A light used to create an effect rather than supply overall illumination to the subject, e.g. a backlight on the hair of a model.

Emulsion Light sensitive layer applied to paper or film.

Enlarger Equipment for projecting the image from the film onto light sensitive photographic paper in the darkroom.

Evaluate Estimate the value or quality of a piece of work.

Exposure A controlled amount of light used to create an image on a light sensitive material. This is a combination of the amount of time the material is subjected to and the aperture of the lens used to form the image.

F-numbers These are the numbers given to the sizes of the apertures in the lens. The f.numbers are standard on all lenses of all makes. The largest number corresponds to the smallest aperture and viceversa, e.g. f.16 is a small aperture on a lens for a 35mm SLR camera and lets less light pass through it than f.4, which is a relatively large aperture.

Fast film A film that is very light sensitive compared to other films. A fast film is one with a high ISO number, e.g. 400 and can be used where the light is not very bright.

Fill light This is a light source which is used to soften the shadows created by the main light.

Film speed A precise number given to film that will indicate how light sensitive it is. See 'Fast film' and 'Slow film'.

Filter A thin sheet of glass or plastic placed in front of the camera lens to alter the appearance of the image.

Fixer A chemical used to make the image stable and permanent after it has been developed.

Focal length A measurement in millimetres between a sharp image and the lens. A lens with a long focal length, e.g. 150mm, will magnify a small part of the subject onto the picture frame. A lens with a short focal length, e.g. 28mm, will record a larger part of the subject in the picture frame.

Focal point The point in the image to which the viewer's attention is attracted.

Focusing The action of creating a sharp image by adjusting either the distance of the lens from the film or altering the position of lens elements.

Fogging Accidental exposure of film or paper to white light.

Format This refers either to the image area created by the camera or the orientation of the image itself, e.g. landscape format refers to an image which has its longest side positioned horizontally.

Frame the act of composing an image. See 'Composition'.

Golden section A classical method for obtaining a good composition.

Grain Tiny particles of the light sensitive crystals which make up the film emulsion. Fast films have larger grain than slow films. Focus finders are used to magnify the projected image so that the grain can be seen and an accurate focus obtained.

Hard light A light source which appears small to the human eye and produces directional light giving well-defined shadows, e.g. direct sunlight or a naked light bulb will give 'hard light'.

Horizontal A line that is parallel to the horizon.

ISO These letters are accompanied by a series of numbers that refer to the speed of the film. They stand for 'International Standards Organisation'.

Juxtapose Place side by side.

Infrared film A film which is sensitive to wavelengths of light longer than 720nm but which are invisible to the human eye.

LED. Small lights inside the viewfinders of some cameras are 'Light Emitting Diodes' or LEDs for short. They may inform the photographer of the settings of the aperture, shutter speed and light meter reading without the photographer having to remove the camera from the eye.

Lens An optical device usually made from glass that focuses light rays to form an image on a surface.

Light meter A device that measures the intensity of light so that the correct exposure for the film can be obtained. A light meter is an integral part of most SLR cameras but may be purchased as a seperate hand held unit.

Lith film A high contrast film that is not sensitive to red light.

Masking The process by which light or chemicals are prevented from reaching a light sensitive surface.

Multiple exposure A technique where several exposures are made onto the same frame of film or piece of paper.

Negative An image on film or paper where the tones are reversed, e.g. dark tones are recorded as light tones and viceversa.

Neutral density filter A filter which reduces the amount of light reaching the film. This enables the photographer to choose large apertures in bright conditions or extend exposure times for special effects.

Opaque Something that does not allow light to pass through it.

Open up A term referring to the action of increasing the lens aperture to let more light reach the film.

Panning The act of moving the camera to follow the subject.

Perspective The illusion of depth and distance in a two-dimensional print. Horizontal lines converge towards the same point on the horizon line.

Polarising filter A filter used to remove unwanted reflections from glass and water. A polarising filter may also increase the colour saturation and darken blue skies in certain lighting conditions.

Pushing film The film speed on the camera's dial is increased to a higher number for the entire film. This enables the film to be used in low light conditions. The film must be developed for longer to compensate for the underexposure.

Reflector A reflective surface used in place of a fill light to soften harsh shadows.

Refraction The change in direction of light as it passes through a transparent surface at an angle.

Resin coated paper Plastic coated photographic paper which allows rapid processing and drying.

Reticulation   Crazed effect which happens when the soft emulsion is subjected to extreme temperature changes.

Rule of thirds   An imaginary grid which divides the frame into three equal sections vertically and horizontally. The lines and intersections of this grid are used to design a pleasing composition.

Safelight   A light in a darkroom with a colour and intensity which will not fog certain photographic emulsions.

Scale   A ratio of size. The relationship of size between the image and the real subject, e.g. something that appears as one centimetre in the image that is 10 centimetres in real life is said to have a scale of 1:10.

Self-timer   A device which delays the action of the shutter release. This is often used to enable the photographer to be in the picture or when no cable release is available for a long exposure.

Sharp   An image is said to be sharp when it is in focus.

Shutter   A mechanism that controls the accurate duration of the exposure.

Silhouette   The outline of a subject seen against a bright background.

Skylight filter   A filter used to reduce or eliminate the blue haze seen in landscapes. The filter does not affect the exposure so it is often used to protect the lens from being scratched.

Slow film   A film that is not very sensitive to light when compared to other films with a higher ISO number. The advantage of using a slow film is its smaller grain size.

SLR camera   Single Lens Reflex camera. The image in the viewfinder is the same image that the film will see. This image prior to taking the shot is viewed via a mirror which moves out of the way when the shutter release is pressed.

Soft light   This is another way of describing diffused light which comes from a broad light source and creates shadows that are not clearly defined.

Solarisation   This is an effect created by briefly exposing photographic film or paper to white light during the processing procedure.

Standard lens   This is a 50mm lens on a 35mm SLR camera. This lens recreates an image of a scene that is close in magnification to how the human eye would see it.

Stop bath   A chemical used to stop the action of the developer and prolong the life of the fixer.

Stop down   A term referring to the action of decreasing the aperture of the lens to reduce the amount of light reaching the film.

Surreal images   Images that are constructed from the imagination rather than from reality.

Sync lead   A lead from the camera to the flash unit which synchronises the firing of the flash and the opening of the shutter.

Sync speed   The fastest shutter speed possible when using a flash unit with a 35mm SLR camera. On most cameras this is about 1/60th of a second. If the sync speed of the camera is exceeded the image will not be fully exposed as the shutter will not be fully open when the flash fires. The sync speed on most cameras is indicated by a different colour or a lightning symbol.

Telephoto lens   A lens for a 35mm SLR camera with a focal length which is greater than 50mm. These lenses are often used to photograph distant subjects which the photographer is unable to get close to. They are also used to flatten perspective and decrease the depth of field. They are most often used by sports and wildlife photographers and are also widely used to take portraits.

Test strip A strip of photographic paper used to gauge the intensity of light coming from the enlarger lens so that an accurate exposure can be made.

Theme A set of images with a unifying idea.

Tinting Applying colour in the form of paint, inks or dyes to an image.

Tone A tint of colour or shade of grey.

Tone elimination A technique used to remove specific tones from an image.

Toning A chemical process to replace the black image with a coloured one.

Transparent Something that allows light to pass through it.

TTL metering Through the lens metering. This is a convenient way to measure the brightness of a scene as the meter is behind the camera lens.

Triptych A series of three images where the images either side relate in some way to the central image.

Tungsten light A common type of electric light such as household bulbs and photographic lamps. When used with daylight colour film a blue filter must be used to prevent the resulting prints looking orange.

Uprating film Another term used to describe the action of pushing the film, i.e. the film speed on the camera's dial is increased to a higher number for the entire film. This enables the film to be used in low light conditions. The film must be developed for longer to compensate for the underexposure.

UV filter A filter used to absorb ultraviolet light. The filter appears colourless and is often left on the lens permanently to protect the lens.

Vantage point A position in relation to the subject which enables the photographer to compose a good shot.

Vertical At right angles to the horizontal plane.

Wetting agent A detergent like fluid used in very small quantities which helps prevent water drying on the surface of the film creating stains.

Wide angle lens A lens for a 35mm SLR camera with a focal length shorter than 50mm. These lenses are often used to photograph subjects which require a wide angle of view to get everything in. The photographer may be unable to move further away or want to move closer to increase the perspective. These lenses are often used by landscape photographers and photojournalists who want to create dramatic images through the use of steep perspective.

Visualise To imagine how something will look once it has been completed.

X sync A socket on the camera or flash unit which enables a sync lead to be attached. When this lead is connected the flash will fire in time with the shutter opening.

Zoom lens A lens which can vary its focal length. It is possible to have a wide angle zoom lens, a wide angle to telephoto zoom that covers the standard focal length and a telephoto zoom lens. These lenses are very convenient as they do not have to be changed so often. The drawback of these lenses is that they have relatively small maximum apertures e.g. f.4 or f.5.6 whereas a fixed focal length lens may have an aperture of f.2 or f.2.8.

Zooming This is a technique where the focal length of a zoom lens is altered during a long exposure. The effect creates movement blur which radiates from the centre of the image.

Index